Everyday Mathematics®

The University of Chicago School Mathematics Project

Differentiation Handbook

Grade 4

Mc Graw Hill Education

Chicago, IL • Columbus, OH • New York, NY

The University of Chicago School Mathematics Project (UCSMP)

Max Bell, Director, UCSMP Elementary Materials Component; Director, *Everyday Mathematics* First Edition; James McBride, Director, *Everyday Mathematics* Second Edition; Andy Isaacs, Director, *Everyday Mathematics* Third Edition; Amy Dillard, Associate Director, *Everyday Mathematics* Third Edition; Rachel Malpass McCall, Associate Director, *Everyday Mathematics* Common Core State Standards Edition

Authors
Amy Dillard
Kathleen Pitvorec

Common Core State Standards Edition
Rachel Malpass McCall, Rebecca W. Maxcy, Kathryn M. Rich

Technical Art
Diana Barrie

ELL Consultant
Kathryn B. Chval

Differentiation Assistant
Serena Hohmann

Photo Credits
Cover (l)Tony Hamblin/Frank Lane Picture Agency/CORBIS, (r)Gregory Adams/Lonely Planet Images/Getty Images, (bkgd)John W. Banagan/Iconica/Getty Images; **Back Cover** Gregory Adams/Lonely Planet Images/Getty Images; **9** (l)Barrie Rokeach Alamy, (c)Image DJ/Alamy, (r)Damons Point Light/Alamy; **23 33** The McGraw-Hill Companies; **Icons** (Objective)Brand X Pictures/PunchStock/Getty Images.

Permissions
Carl Sagan quotation 9806 from Dictionary of Quotations Third Edition, Newly Revised, reproduced by kind permission of Wordsworth Editions, LTD.

Deciding to Teach Them All, Tomlinson, C., Educational Leadership 61(2), © 2003, reprinted by permission. The Association for Supervision and Curriculum Development is a worldwide community of educators advocating sound policies and sharing best practices to achieve the success of each learner. To learn more, visit ASCD at www.ascd.org.

Gregory, G., Differentiated Instructional Strategies in Practice, p. 27, © 2003 by Corwin Press Inc., reprinted by permission of Corwin Press Inc.

 This material is based upon work supported by the National Science Foundation under Grant No. ESI-9252984. Any opinions, findings, conclusions, or recommendations expressed in this material are those of the authors and do not necessarily reflect the views of the National Science Foundation.

everyday**math**.com

Send all inquiries to:
McGraw-Hill Education
STEM Learning Solutions Center
P.O. Box 812960
Chicago, IL 60681

ISBN 978-0-07-657647-0
MHID 0-07-657647-7

Printed in the United States of America.

1 2 3 4 5 6 7 8 9 RHR 17 16 15 14 13 12 11

McGraw-Hill is committed to providing instructional materials in Science, Technology, Engineering, and Mathematics (STEM) that give all students a solid foundation, one that prepares them for college and careers in the 21st century.

The McGraw-Hill Companies

Contents

Differentiating Instruction with *Everyday Mathematics*®

Philosophy

> *Differentiation is a philosophy that enables teachers to plan strategically in order to reach the needs of the diverse learners in classrooms today.*
>
> (Gregory 2003, 27)

This handbook is intended as a guide to help you use *Everyday Mathematics* to provide differentiated mathematics instruction. A differentiated classroom is a rich learning environment that provides students with multiple avenues for acquiring content, making sense of ideas, developing skills, and demonstrating what they know.

In this sense, differentiated instruction is synonymous with good teaching. Many experienced teachers differentiate instruction intuitively, making continual adjustments to meet the varying needs of individual students. By adapting instruction, teachers provide all students opportunities to engage in lesson content and to learn.

Though students follow different routes to success and acquire concepts and skills at different times, the philosophy of *Everyday Mathematics* is that all students should be expected to achieve high standards in their mathematics education, reaching the Grade-Level Goals in *Everyday Mathematics* and the benchmarks established in district and state standards.

Everyday Mathematics is an ideal curriculum for differentiating instruction for a variety of reasons. The *Everyday Mathematics* program:

◆ begins with an appreciation of the mathematical sensibilities that students bring with them to the classroom and connects to students' prior interests and experiences;

◆ incorporates predictable routines that help engage students in mathematics and regular practice in a variety of contexts;

◆ provides many opportunities throughout the year for students to acquire, process, and express mathematical concepts in concrete, pictorial, and symbolic ways;

◆ extends student thinking about mathematical ideas through questioning that leads to deepened understandings of concepts;

- incorporates and validates a variety of learning strategies;

- emphasizes the process of problem solving as well as finding solutions;

- provides suggestions for enhancing or supporting students' learning in each lesson;

- encourages collaborative and cooperative groupings in addition to individual and whole-class work;

- facilitates the development and use of mathematical language and promotes academic discourse;

- provides teachers with information about the learning trajectories or paths to achieving Grade-Level Goals;

- highlights opportunities for teachers to assess students in multiple ways over time;

- suggests how students can demonstrate what they know in a variety of ways; and

- encourages students to reflect on their own strengths and weaknesses.

The purpose of this handbook is to provide ideas and strategies for differentiating instruction when using *Everyday Mathematics*. This handbook highlights differentiation that is embedded in the program and also points to features that can be readily adapted for individual students. The information and suggestions will help you use *Everyday Mathematics* to meet the needs of all learners—learners who need support in developing concepts, learners who need support in developing language proficiency, and learners who are ready to extend their mathematical knowledge and skills.

This handbook includes the following:

- a lesson overview to highlight the features that support differentiated instruction;

- general differentiation strategies and ideas for developing vocabulary, playing games, and using Math Boxes, as well as suggestions for how to implement the lessons to differentiate learning effectively;

- specific ideas for differentiating the content of each unit, including suggestions for supporting vocabulary and adjusting the level of games; and

- a variety of masters that can be used to address the needs of individual learners.

A Lesson Overview

Everyday Mathematics lessons are designed to accommodate a wide range of academic abilities and learning styles. This lesson overview highlights some of the strategies and opportunities for differentiating instruction that are incorporated into the lessons.

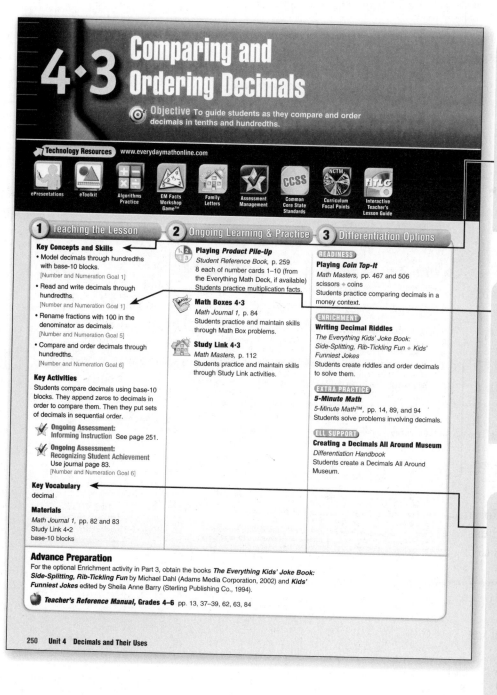

Key Concepts and Skills are identified for each lesson and are linked to Grade-Level Goals. They highlight the variety of mathematics that students may access in the lesson and show that each lesson has significant mathematics content for every student.

Grade-Level Goals are mathematical goals organized by content strand and articulated across grade levels. These goals define a progression of concepts and skills from Kindergarten through Grade 6.

Key Vocabulary consists of words that are new or unfamiliar to students and is consistently highlighted. Students, including English language learners, are encouraged to use this vocabulary in meaningful ways throughout the lesson in order to develop a command of mathematical language.

Mental Math and Reflexes problems range in difficulty, beginning with easier exercises and progressing to more-difficult ones; levels are designated by the symbols ●○○, ●●○, and ●●●. Many of these activities are presented in a "slate, chalk, and eraser" format that engages all students in answering questions and allows the teacher to quickly assess students' understanding.

Math Messages activate and build on students' prior knowledge and create a context for the material to be learned.

Informing Instruction notes suggest how to use observations of students' work to adapt instruction. These notes are designed to help the teacher anticipate and recognize common errors and misconceptions in students' thinking or to alert the teacher to multiple solution strategies or unique insights students may offer.

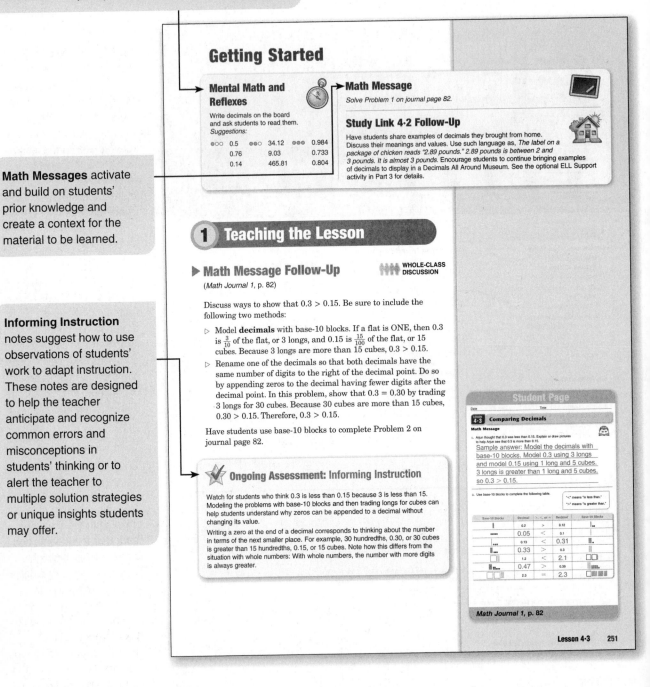

Getting Started

Mental Math and Reflexes

Write decimals on the board and ask students to read them.
Suggestions:

●○○	0.5	●●○ 34.12	●●● 0.984
	0.76	9.03	0.733
	0.14	465.81	0.804

Math Message

Solve Problem 1 on journal page 82.

Study Link 4·2 Follow-Up

Have students share examples of decimals they brought from home. Discuss their meanings and values. Use such language as, *The label on a package of chicken reads "2.89 pounds." 2.89 pounds is between 2 and 3 pounds. It is almost 3 pounds.* Encourage students to continue bringing examples of decimals to display in a Decimals All Around Museum. See the optional ELL Support activity in Part 3 for details.

1 Teaching the Lesson

▶ Math Message Follow-Up

WHOLE-CLASS DISCUSSION

(*Math Journal 1*, p. 82)

Discuss ways to show that 0.3 > 0.15. Be sure to include the following two methods:

▷ Model **decimals** with base-10 blocks. If a flat is ONE, then 0.3 is $\frac{3}{10}$ of the flat, or 3 longs, and 0.15 is $\frac{15}{100}$ of the flat, or 15 cubes. Because 3 longs are more than 15 cubes, 0.3 > 0.15.

▷ Rename one of the decimals so that both decimals have the same number of digits to the right of the decimal point. Do so by appending zeros to the decimal having fewer digits after the decimal point. In this problem, show that 0.3 = 0.30 by trading 3 longs for 30 cubes. Because 30 cubes are more than 15 cubes, 0.30 > 0.15. Therefore, 0.3 > 0.15.

Have students use base-10 blocks to complete Problem 2 on journal page 82.

✔ Ongoing Assessment: Informing Instruction

Watch for students who think 0.3 is less than 0.15 because 3 is less than 15. Modeling the problems with base-10 blocks and then trading longs for cubes can help students understand why zeros can be appended to a decimal without changing its value.

Writing a zero at the end of a decimal corresponds to thinking about the number in terms of the next smaller place. For example, 30 hundredths, 0.30, or 30 cubes is greater than 15 hundredths, 0.15, or 15 cubes. Note how this differs from the situation with whole numbers: With whole numbers, the number with more digits is always greater.

Lesson 4·3 **251**

Recognizing Student Achievement notes highlight specific tasks that can be used for assessment to monitor students' progress toward Grade-Level Goals. The notes identify the expectations for a student who is making adequate progress and point to skills or strategies that some students may be able to demonstrate.

Games played in the classroom, online, and at home provide significant practice in *Everyday Mathematics*. Games are ideal for differentiating instruction as rules and levels of difficulty can be modified easily.

Adjusting the Activity notes include recommendations for tools, visual aids, and other instructional strategies that provide immediate support for exceptional learners. These notes also provide suggestions for open-ended questions to extend students' thinking. Notes labeled "ELL" include suggestions for meeting the needs of English language learners.

Math Boxes are designed to provide distributed practice. Math Boxes routinely revisit recent content to help students build and maintain important concepts and skills. One or two problems on each journal page preview content for the coming unit. Use class performance on these problems as you plan for the coming unit.

Writing/Reasoning prompts are linked to Math Boxes problems. These prompts provide students with opportunities to respond to questions that extend and deepen their mathematical thinking. In addition, these prompts offer regular opportunities for students to communicate their understanding of concepts and skills and their strategies for solving problems.

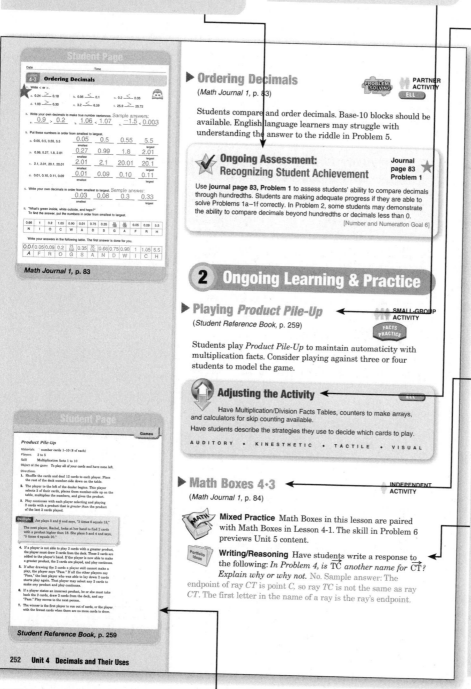

The **Student Reference Book** is a resource for students to use with their teachers, families, and classmates. It includes examples of completed problems similar to those students encounter in class, explanations, illustrations, and game directions. The *Student Reference Book* provides excellent support for all students, including English language learners and their families. At Grades 1 and 2, this book is called **My Reference Book**.

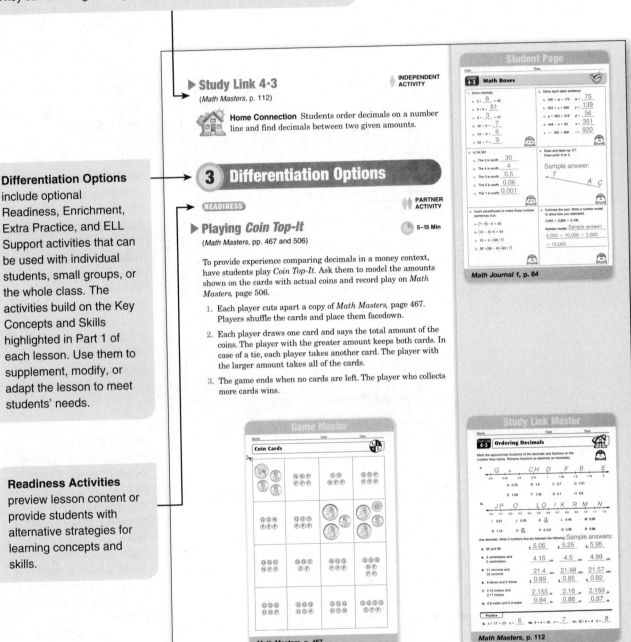

▶ **Study Link 4·3**

(Math Masters, p. 112)

INDEPENDENT ACTIVITY

Home Connection Students order decimals on a number line and find decimals between two given amounts.

3 Differentiation Options

READINESS

PARTNER ACTIVITY

▶ **Playing *Coin Top-It***

5–15 Min

(Math Masters, pp. 467 and 506)

To provide experience comparing decimals in a money context, have students play *Coin Top-It*. Ask them to model the amounts shown on the cards with actual coins and record play on *Math Masters,* page 506.

1. Each player cuts apart a copy of *Math Masters,* page 467. Players shuffle the cards and place them facedown.

2. Each player draws one card and says the total amount of the coins. The player with the greater amount keeps both cards. In case of a tie, each player takes another card. The player with the larger amount takes all of the cards.

3. The game ends when no cards are left. The player who collects more cards wins.

Math Journal 1, p. 84

Math Masters, p. 467

Math Masters, p. 112

Lesson 4·3 253

ENRICHMENT

PARTNER ACTIVITY

15–30 Min

▶ **Writing Decimal Riddles**

Portfolio Ideas

Literature Link To apply students' understanding of decimal concepts, have them write and solve decimal riddles similar to the one on journal page 83. The following books are good sources for riddles:

▷ *The Everything Kids' Joke Book: Side-Splitting, Rib-Tickling Fun* (Everything Kids Series) by Michael Dahl (Adams Media Corporation, 2002)

▷ *Kids' Funniest Jokes,* edited by Sheila Anne Barry (Sterling Publishing Co., 1994)

EXTRA PRACTICE

SMALL-GROUP ACTIVITY

5–15 Min

▶ *5-Minute Math*

To offer students more experience with decimals, see *5-Minute Math,* pages 14, 89, and 94.

ELL SUPPORT

SMALL-GROUP ACTIVITY

15–30 Min

▶ **Creating a Decimals All Around Museum**

(*Differentiation Handbook*)

To provide language support for decimals, have students create a Decimals All Around Museum. See the *Differentiation Handbook* for additional information.

Ask students to read the numbers and describe some of the ways that decimals are used in the museum; for example, what the numbers mean, the different categories of uses, or the units attached to the decimals.

Features for Differentiating in Everyday Mathematics

General Differentiation Strategies

> *All tasks should respect each learner. Every student deserves work that is focused on the essential knowledge, understanding, and skills targeted for the lesson. Every student should be required to think at a high level and should find his or her work interesting and powerful.*
>
> (Tomlinson 2003, 61, 2: 9)

Each *Everyday Mathematics* lesson focuses on a range of mathematical concepts and skills. The most prominent of these are highlighted in the *Key Concepts and Skills* section at the beginning of the lesson. Planning for differentiated instruction involves analyzing which Key Concepts and Skills are appropriate as learning objectives for individual students and then supporting, emphasizing, and enhancing these concepts and skills when teaching the lesson.

Examples of some of the instructional strategies incorporated into *Everyday Mathematics* lessons are described here. These strategies will help you support, emphasize, and enhance lesson content to ensure that all students, including English language learners, are engaged in the mathematics at their appropriate developmental level.

Framing the Lesson
Lesson introductions set the stage and support learning by mentally preparing students for the content of the lesson or by activating prior knowledge. For example, you might begin a geometry lesson with one of the following:

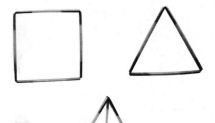

◆ Remind students that they were working on 2-dimensional shapes in the last lesson. Have them discuss what they remember about 2-dimensional shapes.

◆ Tell students that today they are going to build geometric shapes using straws. Ask: *What are some things you know about shapes that will help you with this activity?*

Providing Wait Time

Lessons consist of whole-class, small-group, partner, and independent work. During the whole-class portion of a lesson, allow time for students to think and process information before eliciting answers to questions posed. Waiting even a few seconds for an answer will help many students process information and, in turn, participate more fully in class discussions.

Wait time is also beneficial when you pose Mental Math and Reflexes problems. Encourage students to stop and think before they write on their slates and show their answers. Consider displaying the three steps on a poster. Establish a routine by pointing to the steps in sequence, pausing at each for several seconds.

Establish a routine using Mental Math and Reflexes in which students Think, Write, and Show.

Making Connections to Everyday Life

Lessons offer regular opportunities to build on students' everyday life by helping them make connections between their everyday experiences and new mathematics concepts and skills.

Students build an Array Museum to display examples of arrays found in everyday life. Arrays are closely related to equal-groups situations. If the equal groups are arranged in rows and columns, then a rectangular array is formed. As with equal-group situations, arrays can lead to either multiplication or division problems.

Modeling Concretely

Everyday Mathematics lessons frequently include the use of manipulatives. Make them easily available at all times and for all students. Modeling concretely not only makes lesson content more accessible for some students, but it can also deepen all students' understanding of concepts and skills.

◆ Have pattern blocks available so students can model fraction computation problems such as $\frac{2}{3} + \frac{1}{6} = $ ——.

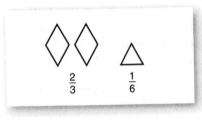

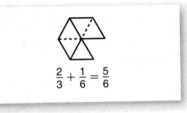

$$\frac{2}{3} + \frac{1}{6} = \frac{5}{6}$$

Model the fractions to be added with pattern blocks.

Combine the blocks to show the sum.

◆ Have stick-on notes available for making line plots and finding the median.

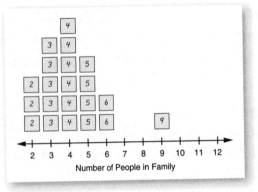

Use stick-on notes to make a line plot showing the number of people students have in their families.

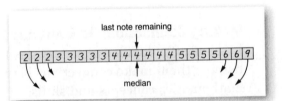

Find the median by lining up the stick-on notes and removing notes two at a time, one from each end, until only one or two notes are left.

◆ Have base-10 blocks available so students can model place value, addition, and subtraction.

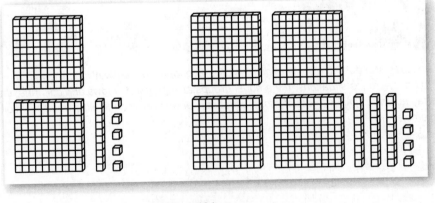

$$215 + 434 = \text{\underline{\hspace{2cm}}}$$

Modeling Visually

Classrooms tend to be highly verbal places, and this can be overwhelming for some students. Simple chalkboard drawings, diagrams, and other visual representations can help students make sense of the flow of words around them and can also help them connect words to the actual items.

◆ Use pictures to model even and odd numbers.

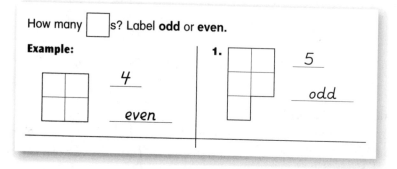

◆ Use arrays to model square numbers.

Complete the facts.

1. $1 \times 1 =$ _1_
2. $2 \times 2 =$ _4_
3. $3 \times 3 =$ _9_
4. $4 \times 4 =$ _16_
5. $5 \times 5 =$ _25_
6. $6 \times 6 =$ _36_
7. $7 \times 7 =$ _49_
8. $8 \times 8 =$ _64_
9. $9 \times 9 =$ _81_
10. $10 \times 10 =$ _100_

◆ Use a number line to visually model division by a fraction.

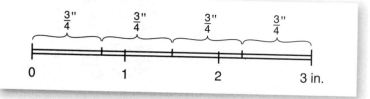

To illustrate division of a whole number by a fraction, students partition a 3-inch segment into equal $\frac{3}{4}$-inch segments. Students ask the question, "How many $\frac{3}{4}$-inch segments are in 3 inches?" They answer the question by counting the number of line segments; in this case, there are 4 equal segments. Students then write the number sentence, $3 \div \frac{3}{4} = 4$.

Modeling Physically

Lessons suggest ways to have students demonstrate concepts and skills with gestures or movements. This strategy helps many students better understand and retain the concept or skill.

◆ Have students model the concept of *parallel* by holding their arms in front of them, parallel to each other.

A physical model for parallel line segments

◆ Have students model addition and subtraction problems by moving their fingers on number lines or number grids. A number-grid master can be found on page 146 of this handbook.

-9	-8	-7	-6	-5	-4	-3	-2	-1	0
1	2	3	4	5	6	7	8	9	10
11	12	13	14	15	16	17	18	19	20
21	22	23	24	25	26	27	28	29	30
31	32	33	34	35	36	37	38	39	40
41	42	43	44	45	46	47	48	49	50
51	52	53	54	55	56	57	58	59	60
61	62	63	64	65	66	67	68	69	70
71	72	73	74	75	76	77	78	79	80
81	82	83	84	85	86	87	88	89	90
91	92	93	94	95	96	97	98	99	100
101	102	103	104	105	106	107	108	109	110

$15 + 33 =$ _____

◆ Have students skip count on a calculator while doing a class count. This strategy reinforces counting visually by showing the numbers while at the same time physically engaging students.

Program a TI-15 calculator to count by tenths. Clear the calculator. Enter **Op1** **+** *0.1* **Op1** *0* **Op1** *and repeatedly enter* **Op1** *without clearing the calculator.*

Providing Organizational Tools

Lessons provide a variety of tools to help students organize their thinking. Have students use diagrams, tables, charts, and graphs when these materials are included in lessons and as appropriate. This is another way to make the lesson content more accessible for some students while at the same time deepening other students' understanding of concepts and skills.

◆ Have students use Venn diagrams to compare and contrast properties of numbers, shapes, and so on.

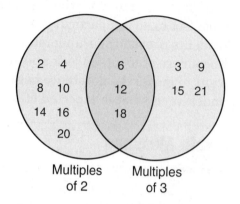

Blank masters for Venn diagrams can be found on pages 147 and 148 of this handbook.

◆ Have students use situation diagrams to model operations. Blank masters for these diagrams can be found on pages 149 and 150 of this handbook.

rows	chairs per row	chairs in all
3	?	15

Adriana set up chairs in her backyard for a play. She had 15 chairs in all. She made 3 rows. How many chairs were in each row?

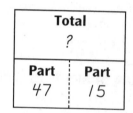

Total	
?	
Part	**Part**
47	15

Malcolm had 47 pennies in a jar in his room. His brother had 15 pennies. How many pennies did they have in all?

◆ Provide students with place-value charts or have them draw their own. Have them write numbers in the charts as dictated, for example, the number that has a 3 in the thousands place, a 2 in the ones place, a 4 in the ten-thousands place, and a 0 everywhere else.

Ten Thousands	Thousands	Hundreds	Tens	Ones
4	3	0	0	2

Engaging Students in Talking about Math

Lessons often suggest discussion prompts or questions that support the development of good communication skills in the context of mathematics. Although finding the correct solution is one important goal, *Everyday Mathematics* lessons also emphasize sharing and comparing solution strategies. This type of "math talk" involves not only explaining what is done (explanation), but also why it is done (reasoning), and why it is correct or incorrect to do it a particular way (justification). These discussions help students deepen their understanding of mathematical concepts and processes. Encourage students to look at other students when they are speaking. You may want to model the difference between hearing and listening to help students understand what is expected of them.

Math Message

Tell whether each number sentence is true or false.

$$28 - 6 + 9 = 31$$
$$28 - 6 + 9 = 13$$

Be ready to defend your answer.

Some students may work the problems from left to right and determine that the first number sentence is true. Other students may decide that the second number sentence is true by first adding 6 and 9 and then subtracting the sum from 28. Others may reason that both sentences could be true, depending on what you do first. This Math Message problem and the resulting discussion serve as an introduction to the use of parentheses in number sentences that involve more than one operation.

Engaging Students in Writing about Math

Journal pages and assessment problems frequently prompt students to explain their thinking and strategies in words, pictures, and diagrams. Writing offers students opportunities to reflect on their thinking and can help you assess their mathematical understandings and communication skills. Exit Slips and Math Logs are ideal places for students to record their thinking.

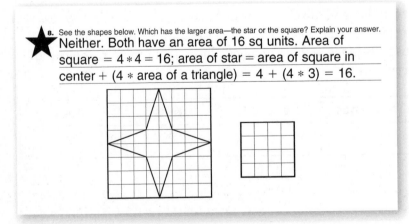

8. See the shapes below. Which has the larger area—the star or the square? Explain your answer.
   Neither. Both have an area of 16 sq units. Area of square = 4 * 4 = 16; area of star = area of square in center + (4 * area of a triangle) = 4 + (4 * 3) = 16.

Students demonstrate their understanding of area by responding to the question on the journal page.

Using Key Concepts and Skills

Each *Everyday Mathematics* lesson provides students with opportunities to explore a variety of mathematics. This variety allows you to target appropriate concepts and skills for individual students.

Shown below are the Key Concepts and Skills in Lesson 3-4 of *Third Grade Everyday Mathematics.*

Key Concepts and Skills
- Use basic facts to find perimeter. [Operations and Computation Goal 1]
- Model polygons with straws; identify and describe polygons. [Geometry Goal 2]
- Measure sides of polygons to the nearest inch. [Measurement and Reference Frames Goal 1]
- Add side lengths to find perimeter. [Measurement and Reference Frames Goal 2]

At the beginning of the lesson, students use straws and connectors to build the polygons in Problems 1 and 2 on journal page 63 and compare the properties of these polygons. Students then work on the journal page.

◆ Modeling and describing polygons and measuring the lengths of the sides may be reasonable skills to target for some students in this lesson. These students might complete only Problems 1 and 2 on journal page 63.

◆ Problems 1 and 2 may be the most important ones for some students to complete. Encourage students to complete these problems first and to finish the remainder of the page if they have time, comparing their answers with one another. Circulate and assist.

◆ Finding perimeters may be a reasonable skill to emphasize for some students. If students completed Problem 4 by drawing a rectangle with a perimeter of 20, ask them to apply their understanding of perimeter by drawing other rectangles on a grid with perimeters of 20. Have students write an explanation of how they can be sure they found all such rectangles.

◆ Some students may need more time to complete all the problems. Have students who do not complete the page during the course of the lesson complete it later as time and experience allow.

Math Journal 1, *page 63, reflects the Key Concepts and Skills in Lesson 3-4 of* Third Grade Everyday Mathematics.

Summarizing the Lesson

Lesson summaries offer students a chance to bring closure to the lesson, reflect on the concepts and skills they have learned, and pose questions they may still have about the lesson content. Exit Slips and Math Logs are ideal places for students to record their reflections about what they learned. For example, a lesson on measurement might close with one of the following:

◆ Have students describe what they learned about standard units of linear measure.

Name	Date

My Exit Slip

I measured my desk. I used crayons and I used my shoes. I got different measures. If I used a ruler I always got the same.

Using an Exit Slip, a student describes what she learned.

◆ Have students record what they know about using a ruler to measure length.

My Exit Slip

When I use a ruler to measure length I have to line up one edge of the line segment with the 0 on the ruler. Sometimes the 0 isn't at the edge.

Using an Exit Slip, a student explains how he uses a ruler.

Vocabulary Development

The most effective way for students, including English language learners, to learn new words is to encounter them repeatedly in meaningful contexts. When the meaning of a new word is understood, real mastery requires using it in conversation and writing. With this principle in mind, *Everyday Mathematics* incorporates many opportunities within the lessons for students to develop vocabulary. *For example:*

◆ Topics and concepts are regularly revisited throughout the program, so students are constantly building on and deepening their understanding of mathematical terms from previous lessons.

◆ Hands-on, interactive, and visual activities in each lesson ensure that new words are introduced in clear, comprehensible ways.

◆ Sharing solutions and explanations, along with cooperative group work, ensures that students have opportunities to use new vocabulary purposefully.

Examples of helpful strategies are described here.

Providing Visual References

Suggest visual references to provide support for the use and development of mathematical language.

◆ Have students underline the names of the pattern blocks with a pencil that is the same color as the corresponding block to help students associate the words with the shapes.

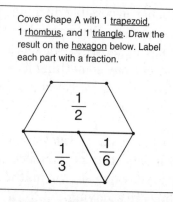

◆ Record the words and the number model for a number story.

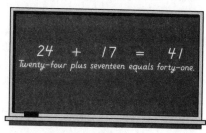

Using the *Student Reference Book*

The *Student Reference Book* is a rich resource for definitions and examples of vocabulary. Teach students how to use the table of contents, index, and glossary to make optimal use of this resource. The *Student Reference Book* is also a good source of illustrations that English language learners often find useful.

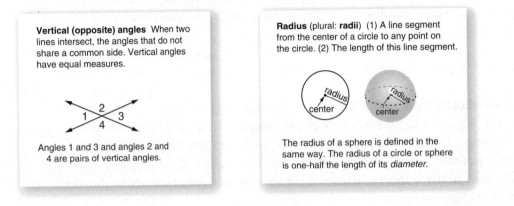

Creating a Language-Rich Environment

Support students' development of their mathematics vocabulary by immersing them in a language-rich environment. Seeing, hearing, and using new terms in meaningful ways will help them navigate through the language-rich mathematics lessons and will support their development of stronger communication skills.

◆ Display new vocabulary on a Math Word Wall. Include illustrations so that students can make sense of the words and use them in their speech and writing.

◆ Use mathematical terminology whenever possible during class discussions. For example, instead of saying, *A square has four corners,* say something like, *A square has four vertices, or corners.*

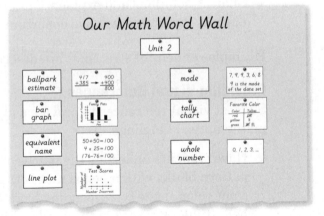

Samples from a Math Word Wall

◆ Post labels in the classroom that will help students connect their everyday lives to the mathematics they are studying.

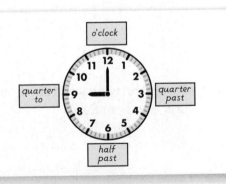

Label your classroom clock.

Recording Key Ideas

On the board or Class Data Pad, record in words, symbols, and pictures the key ideas or key solution steps that students share during class discussions. *For example:*

During a class discussion, one student shares the following strategy when explaining how he solved 31 – 14. "First I took 10 from the 14 and subtracted it from 30 because there is a 30 and a 1 in 31. I had 21 left. Then I counted back 4 more for the 4 I had left from 14. I started with 21, then counted, 20, 19, 18, 17. I was holding up four fingers, so I knew I counted back 4."

Record the steps of a student's strategy on the board.

Clarifying the Meaning of Words

Lessons routinely highlight potentially confusing words and provide suggestions for clarifying their meanings. Such words include those with multiple meanings, such as *power,* and homophones (words that are pronounced the same, but differ in meaning), such as *sum* and *some.*

◆ Discuss the everyday versus the mathematical usage of the word *change*.

◆ Discuss the different meanings of the word *power* in the terms *fact power* and *power of a number.*

◆ Write *some* and *sum* on the board. Discuss and clarify the meaning of each word.

Illustrate the difference between sum *and* some.

Games

Frequent practice is necessary for students to build and maintain strong mental-arithmetic skills and reflexes. There are many opportunities in *Everyday Mathematics* for practice through games. Games are not merely attractive add-ons but an essential component of the *Everyday Mathematics* program and curriculum.

Everyday Mathematics games are important for these reasons:

◆ Games help students develop the ability to think critically and solve problems. The variety of games in *Everyday Mathematics* lays the foundation for increasingly difficult concepts and helps students develop sophisticated solution strategies.

◆ Games provide an effective and interactive way for students to practice and master basic concepts and skills. Practice through playing games not only builds fact and operation skills, but often reinforces other concepts and skills, such as calculator use, money exchange, geometric intuition, and ideas about probability.

◆ Games have advantages over paper-and-pencil drills.

Games	Paper-and-Pencil Drills
Present enjoyable ways to practice skills	Tend toward tedium and monotony
Can be played during free time, lunch and recess, or even at home	Are used only during required class time
Are worksheet-free	Are worksheet-based
Are easily adaptable for a class of students who need to practice a wide range of skills at a variety of levels	Require a variety of worksheets to practice different skills at a variety of levels
Provide immediate insight into students' understanding through their discussions and conversations about mathematics	Result in attempts to understand students' thinking while grading worksheets that are days old

Spend some time learning the *Everyday Mathematics* games so that you understand how much they contribute to students' mathematical progress and can join in the fun.

Using Games in the Classroom

Games can be used in many ways. Consider these ideas for making games both enjoyable and educational for all students:

◆ Establish a routine to provide all students the opportunity to play games at least two or three times each week for a total of about one hour per week. Practice is most effective when it is distributed, so several short practice sessions are preferable to one large block of time.

◆ Establish a routine for playing games as a regular part of your math class rather than as a reward for completing assigned work. It is important that all students have time to play games, especially students who work at a slower pace or who may need more practice than their classmates do. This way, students who need the practice the most will not miss out.

◆ Set up a Games Corner with some of the students' favorite games. Be sure to include all of the gameboards, materials, and game record sheets needed. Consider creating a task card for each game. Encourage students to visit this corner during free time. Change the games menu frequently to correspond with concepts currently taught in your classroom and to offer students additional practice and review of particular skills.

> *LANDMARK SHARK*
> See Student Reference Book
> pages 325 and 326 for instructions.
> *Materials you need:*
> Landmark Shark score sheet
> (Math Masters, page 457)
> Landmark Shark cards
> (Math Masters, page 456)
> Everything Math Deck

Sample task card for Landmark Shark

◆ Establish game stations where students can rotate to a new station about every 15 minutes. Station time can occur at the beginning or the end of a lesson, during the entire mathematics time, or when a substitute teacher is in the classroom. Consider asking parent volunteers to assist at stations. Provide parents with game directions ahead of time so they are familiar with the rules and with the concepts or skills practiced.

◆ Monitor students when they play games. Ask students to explain the concept or skill they are practicing or describe strategies they are using.

◆ Consider students' strengths carefully when pairing or grouping them. Group students so that they can support one another's learning.

- If it is a new game, consider pairing students who will readily understand and implement the rules with students who may need assistance learning the game.
- If a familiar game can be played at a variety of levels, consider pairing students who are working at the same level.

◆ Have students complete game record sheets so they are accountable for the work they do. Alternatively, have students complete Exit Slips summarizing the concepts or skills they practiced.

Name		Date	Time	

Angle Tangle Record Sheet

Round	Angle	Estimated measure	Actual measure	Score
1		120°	108°	12
2		75°	86°	11
3		40°	44°	4
4		60°	69°	9
5		135°	123°	12
			Total Score	48

Fourth- through sixth-grade students complete an Angle Tangle *record sheet.*

Modifying Games

Games are easily adapted to meet a variety of practice needs. For example, you can engage all students in the same game at a variety of levels. The modification strategies suggested below can be used for most games included in *Everyday Mathematics*. For specific variations, see the game adaptations in the unit-specific section of this handbook beginning on page 47.

◆ Modify the level of difficulty of games by targeting a certain range of numbers for students working at different levels. Because numbers in most games are generated randomly, you can modify blank spinners, decrease or increase the number of dice, roll polyhedral dice, or use specific sets of number cards.

Two 6-sided dice for regular game play

Two 8-sided dice to increase the range of numbers

One 10-sided die (0 through 9) to decrease the range of numbers

◆ Modify the level of difficulty of games by encouraging students to play a mental-math version of a game in which students would normally use paper and pencil to calculate scores.

◆ See whether variations of a game are available so you can target different concepts or skills or different levels for students appropriately. Many *Everyday Mathematics* games provide a range of practice options by including a variety of gameboards or rules.

Hitting Table	
1 to 10 Facts	
1 to 21	Out
24 to 45	Single (1 base)
48 to 70	Double (2 bases)
72 to 81	Triple (3 bases)
90 to 100	Home Run (4 bases)

Hitting Table	
10 * 10s Game	
100 to 2,000	Out
2,100 to 4,000	Single (1 base)
4,200 to 5,400	Double (2 bases)
5,600 to 6,400	Triple (3 bases)
7,200 to 8,100	Home Run (4 bases)

Versions of Baseball Multiplication *provide practice with different levels of multiplication skills, for example, facts for 1 through 10 and extended facts.*

◆ Have slates on hand for students to draw pictures as they work through problems.

- Make various manipulatives, such as coins and bills, base-10 blocks, and counters available to provide concrete models for practicing concepts and skills.

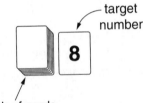

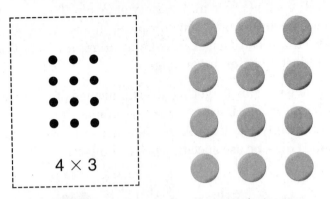

To reinforce multiplication concepts, have students use counters to build the array shown on an Array Bingo card.

- Introduce game-specific or mathematical vocabulary with visual cues, such as writing the terms on the board, as well as auditory support, such as having the class repeat the word aloud as a group. Use new vocabulary consistently and be careful to avoid interchanging or substituting synonymous terms, which can cause confusion for some students.

Target number is a term frequently used in game play.

- Modify the difficulty of games involving target numbers by limiting the numbers that students use. As students gain proficiency, provide larger numbers. Try this with decimals also.

Students playing Hit the Target agree on a 2-digit multiple of 10 as the target number for each round. Players then select a starting number and use their calculators to add or subtract to change the starting number to the target number. You can modify the game by suggesting that students choose as the target number a multiple of 10 less than or equal to 40 or a 3- or 4-digit multiple of 100.

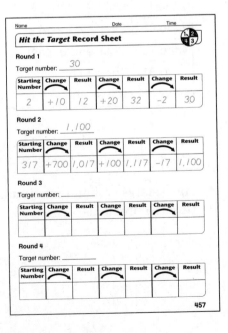

◆ Provide tools such as Addition/Subtraction or Multiplication/Division Facts Tables or calculators for students to check facts quickly and assist them in playing games that require fluency with facts they are learning.

*,/	1	2	3	4	5	6	7	8	9	10	11	12
1	1	2	3	4	5	6	7	8	9	10	11	12
2	2	4	6	8	10	12	14	16	18	20	22	24
3	3	6	9	12	15	18	21	24	27	30	33	36
4	4	8	12	16	20	24	28	32	36	40	44	48
5	5	10	15	20	25	30	35	40	45	50	55	60
6	6	12	18	24	30	36	42	48	54	60	66	72
7	7	14	21	28	35	42	49	56	63	70	77	84
8	8	16	24	32	40	48	56	64	72	80	88	96
9	9	18	27	36	45	54	63	72	81	90	99	108
10	10	20	30	40	50	60	70	80	90	100	110	120
11	11	22	33	44	55	66	77	88	99	110	121	132
12	12	24	36	48	60	72	84	96	108	120	132	144

Multiplication/Division Facts Table

◆ Use illustrations to depict game directions. Create illustrations before introducing the game or during class discussion while introducing the game. Alternatively, have students create the illustrations after they have played the game. Students can refer to the illustrated instructions each time the game is revisited.

Deal 5 cards to each player.

◆ Encourage questions and discussion during games so students can use new vocabulary.

Number Top-It Mat (7-Digit)

	Millions	Hundred Thousands	Ten Thousands	Thousands	Hundreds	Tens	Ones
Andy	7	6	4	5	2	0	1
Barb	4	9	7	3	5	2	4

Students playing Number Top-It *use randomly generated digits to build the largest number possible. Encourage students to discuss and compare strategies for deciding where to place the digits.*

Math Boxes

In *Everyday Mathematics,* Math Boxes are one of the main components for reviewing and maintaining skills. Math Boxes are not intended to reinforce the content of the lesson in which they appear. Rather, they provide continuous distributed practice of concepts and skills targeted in the Grade-Level Goals. It is not necessary for students to complete the Math Boxes page on the same day the lesson is taught, but it is important that the problems for each lesson are completed.

Several features of Math Boxes pages make them useful for differentiating instruction:

◆ Math Boxes in most lessons are linked with Math Boxes in one or two other lessons so that they have similar problems. Because linked Math Boxes pages target the same concepts and skills, they may be useful as extra practice tools.

◆ Writing/Reasoning prompts in the *Teacher's Lesson Guide* provide students with opportunities to respond to questions that extend and deepen their mathematical thinking. Using these prompts, students communicate their understanding of concepts and skills and their strategies for solving problems.

◆ Many Math Boxes problems include an icon for the *Student Reference Book.* This cue tells students where they can find help for completing the problems.

◆ One or two problems on each Math Boxes page preview content from the coming unit. Use these problems identified in the *Teacher's Lesson Guide* to assess student performance and to build your differentiation plan.

◆ The multiple-choice format of some problems provides students with an opportunity to answer questions in a standardized-test format. The choices include *distractors* that represent common student errors. Use the incorrect answers to identify and address students' needs.

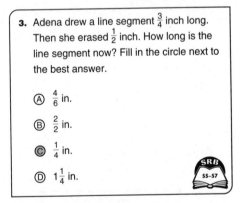

3. Adena drew a line segment $\frac{3}{4}$ inch long. Then she erased $\frac{1}{2}$ inch. How long is the line segment now? Fill in the circle next to the best answer.

Ⓐ $\frac{4}{6}$ in.

Ⓑ $\frac{2}{2}$ in.

Ⓒ $\frac{1}{4}$ in.

Ⓓ $1\frac{1}{4}$ in.

Students choosing $\frac{2}{2}$ in. may have incorrectly subtracted the numerators and denominators. Incorrect algorithm: $\frac{3}{4} - \frac{1}{2} = \frac{3-1}{4-2} = \frac{2}{2}$.

Using Math Boxes in the Classroom

Math Boxes can be used in many ways. Consider these ideas for making Math Boxes a productive learning experience for all students:

◆ Create a cardstock template that allows students to focus on only one problem at a time. Or, have students use stick-on notes to cover all but one problem.

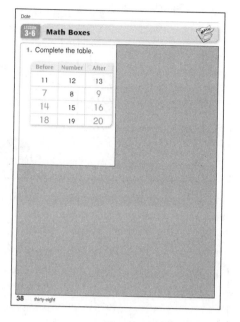
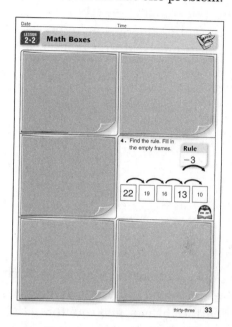

Students can focus on one Math Boxes problem at a time.

◆ Identify the problem or problems that are essential. Encourage students to complete these problems first. Suggest that students who finish a task early use their spare time to complete any unfinished Math Boxes problems. Consider providing time in your weekly schedule so that all students have the opportunity to complete unfinished Math Boxes problems.

◆ Have students complete the problems independently. Then have them form small groups and share their answers and explanations. As an alternative, ask students to complete the problems cooperatively even though the lesson indicates independent work.

◆ Divide the class into groups. Have each group solve one of the Math Boxes problems. "Jigsaw" to form new groups. Each of the new groups now has one student from each of the original groups. Each student in the new group is an expert on one of the problems. The expert explains the problem to the other students in the group.

◆ Have students complete Math Boxes pages as part of their daily morning routine. Math Boxes are one of the components of a lesson that lends itself to being completed outside of regular math time.

Modifying Math Boxes

The strategies suggested here can be used for many types of Math Boxes problems. The same types of modifications can be made to other journal pages as well.

◆ Modify the range of the numbers or ask students to record measurements to a more-precise or less-precise degree of accuracy to focus on a different level of a concept or skill.

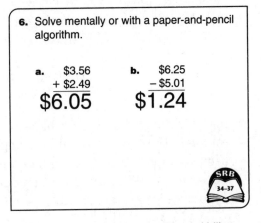

To practice extended facts instead of basic facts, have students attach a zero to each number in the "What's My Rule?" table.

◆ Make various manipulatives, such as coins and bills, base-10 blocks, pattern blocks, and counters, available to provide concrete models for practicing concepts and skills.

Encourage students to use coins and bills to help them solve decimal addition and subtraction problems.

◆ Have tools, such as number grids, place-value charts, calculators, or fact tables, available to help students solve problems.

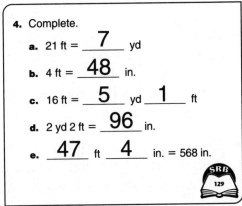

Some students may be able to solve Problems a–c mentally but choose to use a calculator with Problems d and e.

Creating Your Own Math Boxes

Occasionally, you may want to create your own Math Boxes page for practice or assessment purposes. There are blank masters on pages 134–139 of this handbook to serve this purpose. Consider the following ideas while designing pages for your class or individual students:

♦ Create a set of problems that focuses on a single concept or skill that students need to review, but address it in a variety of contexts. For example, focus on addition through number stories, facts problems, "What's My Rule?" problems, or skip counting. Because each Math Boxes page in the journal includes a variety of problems, each one targeting a different concept or skill, this strategy can help students who struggle with these transitions.

♦ Create a Math Boxes page that links with a set of Math Boxes pages in the journal. Tailor the numbers to meet the individual needs of students.

♦ Create a set of extra-practice problems in which all cells focus on concepts or skills from a particular lesson.

♦ Adapt *5-Minute Math* problems that address concepts or skills students need to review.

♦ Create a page in which each problem targets a specific concept or skill. Use the page for one week, each day replacing the numbers in the problems with new numbers.

♦ Use the templates of routines found on pages 136–139 of this handbook to create Math Boxes pages for students to complete. Fill in some of the numbers for each routine. Or, have students create the Math Boxes for classmates to complete. For more information about each of these routines, see the *Teacher's Reference Manual.*

♦ Have students write number stories in each cell of a template. Specify which operation should be the focus of each problem. Have students exchange Math Boxes pages and solve one another's problems.

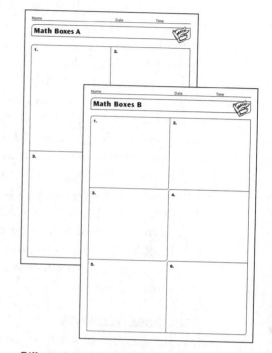

Differentiation Masters, pages 134 and 135, are templates for blank Math Boxes pages for four or six problems.

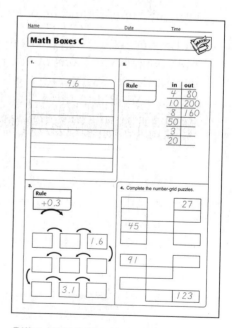

Differentiation Masters, page 136, is a master for a blank Math Boxes page that includes a variety of routines.

Using Part 3 of the Lesson

As written, *Everyday Mathematics* lessons engage a wide range of learners and support the development of mathematics concepts and skills at the highest possible level. There are times, however, when teachers still need to be flexible in implementing lessons. To address the individual needs of students, Part 3 of each lesson, Differentiation Options, provides additional resources beyond the scope of what is included in Part 1. The activities suggested are optional, intended to support rather than to replace lesson content. Many of the activities, designed so students can work with partners or small groups, are ideally suited for station work. Based on your professional judgment and assessments, determine when students might benefit from these activities. For each unit, use the master found on page 152 of this handbook to plan how you will use the Part 3 activities with the whole class, small groups, or individual students.

Part 3 Planning Master

Lesson	Readiness	Enrichment	Extra Practice	ELL Support
1–1		Whole class		Carlos Andres
1–2	Abby Conner Jamal	Chantel Eric	Matt Amy	
1–3			Cheryl DeAndre Melissa	Melanie
1–4	Toya Takako Kevin		Whole class	
1–5		Isabel Leon		Whole class
1–6	Whole class		Whole class	
1–7	Katherine Aman	Jayne	Hannah Abby Tom	Dmitry Carlos

Readiness Activities

Readiness activities introduce or develop the lesson content to support students as they work with the Key Concepts and Skills. Use Readiness activities with some or all students before teaching the lesson to preview the content so students are better prepared to engage in lesson activities. As an alternative, use Readiness activities at the completion of lesson activities to solidify students' understanding of lesson content.

In Lesson 3-1 of *Fourth Grade Everyday Mathematics,* students discuss problems in which one quantity depends on another. They illustrate this kind of relationship between pairs of numbers with a function machine and a "What's My Rule?" table. The following is the Readiness activity for the lesson.

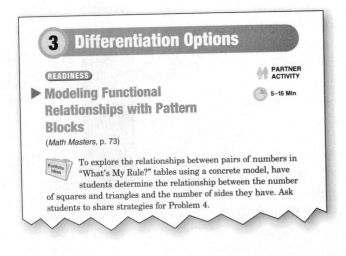

3 Differentiation Options

(READINESS)

► **Modeling Functional Relationships with Pattern Blocks**

(*Math Masters*, p. 73)

PARTNER ACTIVITY

5–15 Min

Portfolio Ideas

To explore the relationships between pairs of numbers in "What's My Rule?" tables using a concrete model, have students determine the relationship between the number of squares and triangles and the number of sides they have. Ask students to share strategies for Problem 4.

Name _____ Date _____ Time _____

LESSON 3-1 **"What's My Rule?" Polygon Sides**

1. Use square pattern blocks to help you complete the table.

Number of Squares	Number of Sides
1	4
2	8
3	12
5	20
7	28
8	32

2. Suppose there are 12 squares. Explain how to find the number of sides without counting.
Sample answer: Multiply 12 squares by 4 sides. This equals 48 sides. ($12 \times 4 = 48$)

3. Use triangle pattern blocks to help you complete the table.

Number of Triangles	Number of Sides
1	3
2	6
5	15
4	12
3	9
6	18

4. Suppose there are 30 sides. Explain how to find the number of triangles without counting.
Sample answer: Divide 30 sides by 3. This equals 10 triangles. ($30 \div 3 = 10$)

73

Enrichment Activities

Enrichment activities provide ways for students to apply or further explore Key Concepts and Skills emphasized in the lesson. Use the activities with some or all students after they have completed the lesson activities.

In Lesson 8-4 of *Second Grade Everyday Mathematics,* students explore the concept of equivalent fractions by matching fractional parts of circles. The following is the Enrichment activity for the lesson.

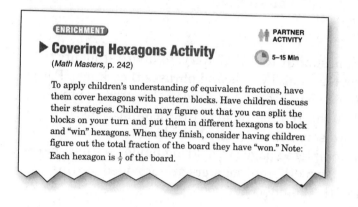

ENRICHMENT

▶ **Covering Hexagons Activity**
(Math Masters, p. 242)

👥 **PARTNER ACTIVITY**

🕐 **5–15 Min**

To apply children's understanding of equivalent fractions, have them cover hexagons with pattern blocks. Have children discuss their strategies. Children may figure out that you can split the blocks on your turn and put them in different hexagons to block and "win" hexagons. When they finish, consider having children figure out the total fraction of the board they have "won." Note: Each hexagon is $\frac{1}{7}$ of the board.

Extra Practice Activities

These activities provide students with additional practice opportunities related to the content of the lesson. There are three main categories for extra practice activities—practice pages, games, and *5-Minute Math* problems.

The *5-Minute Math* problems are grouped by level of difficulty, which allows you to choose problems to meet the needs of your entire class or small groups of students. The activities can also serve as a catalyst for your own or students' problems and ideas.

5-Minute Math activities do the following:

◆ provide reinforcement and continuous review of Grade-Level Goals;

◆ provide practice with mental arithmetic and logical thinking activities;

◆ give students additional opportunities to think and talk about mathematics and to try out new ideas by themselves or with their teachers and classmates; and

◆ promote the process of solving problems, so in the long run, students become more willing to risk sharing their thoughts and their solution strategies with classmates rather than focus on getting quick answers.

Support for English Language Learners

The activities in the ELL Support section are designed to promote development of language related to Key Concepts and Skills. Several vocabulary routines are established early in each grade and are revisited throughout the year.

It is important to note that English language learners should not be restricted solely to ELL Support activities. Often the Readiness and Enrichment activities are ideally suited to enhance the mathematical content for this population of students. Likewise, although the activities described in this section are extremely helpful for English language learners, this kind of work enriches the vocabulary development of all students.

Math Word Bank

Use a Math Word Bank, which is similar to a dictionary, to invite students to make connections between new terms and words and phrases they know. For each entry, have students make a visual representation of the word or phrase and list three related terms that will remind them of the meaning. Have English language learners record some of the related words in their own language. Have students keep completed pages in a 3-ring binder so that they may refer to them as necessary. Two different masters are provided for this routine on pages 140 and 141 of this handbook.

In Lesson 2-1 of *Sixth Grade Everyday Mathematics,* students explore reading and writing large numbers. The following is the ELL Support activity for this lesson.

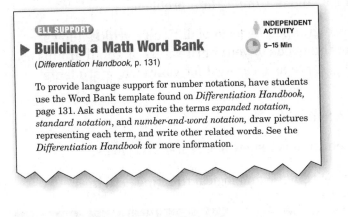

ELL SUPPORT

▶ **Building a Math Word Bank**

(*Differentiation Handbook*, p. 131)

INDEPENDENT ACTIVITY
5–15 Min

To provide language support for number notations, have students use the Word Bank template found on *Differentiation Handbook,* page 131. Ask students to write the terms *expanded notation, standard notation,* and *number-and-word notation,* draw pictures representing each term, and write other related words. See the *Differentiation Handbook* for more information.

Museums

In *Everyday Mathematics,* museums help students connect the mathematics they are studying with their everyday lives. A museum is simply a collection of objects, pictures, or numbers that illustrates or incorporates mathematical concepts related to the lessons. Museums provide opportunities for students to explore and discuss new mathematical ideas. If several English language learners speak the same language, have them take a minute to discuss museums in their own language first and then share in English as they are able.

In Lesson 5-7 of *Second Grade Everyday Mathematics,* students construct pyramids using straws and connectors. They discuss the properties of the pyramids they have built. The following is the ELL Support activity for this lesson.

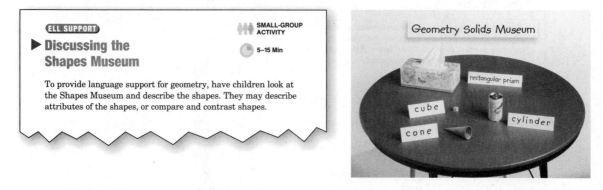

Student- and Teacher-Made Posters

Sometimes a unit focuses on a topic that introduces potentially confusing content, for example, a great deal of new vocabulary, many steps in a problem-solving process, or several strategies for solving a problem. Providing students with a poster to use as a reference or having them create their own posters can help them make sense of such complex content.

In Unit 4 of *Fifth Grade Everyday Mathematics,* students review long-division algorithms. The following is the ELL Support activity for Lesson 4-5 in this unit.

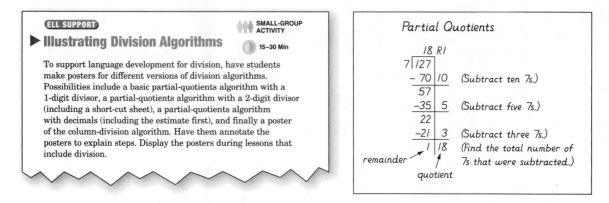

Everyday Mathematics also includes posters, such as the Probability Meter Poster. When you use the *Everyday Mathematics* posters with English language learners, you may display both the English and Spanish versions simultaneously or only the English version.

Graphic Organizers

Students find it easier to learn and retain new words if they connect the new words to their existing vocabulary. Graphic organizers are organizational tools for making connections more explicit and for helping students gain a deeper understanding of a concept.

In Lesson 6-10 of *Second Grade Everyday Mathematics,* students explore multiplication and division and the relationships between these operations. The following is the ELL Support activity for this lesson.

ELL SUPPORT

SMALL-GROUP ACTIVITY
5–15 Min

▶ **Writing Multiplication and Division Phrases**

To provide language support for understanding multiplication and division phrases, have children draw and label a table with three columns on chart paper as shown below. Ask children to identify words or phrases associated with multiplication and division. Write their responses in the table.

Multiplication	Both Multiplication and Division	Division
Addition	Equal Groups	Subtraction
Packages of objects	Arrays	Share
Multiples	Fact Families	How many groups are there?
Skip Counting	Fact Triangles	How many in each group?
All Together		Remaining

In Lesson 8-7 of *Sixth Grade Everyday Mathematics,* students use ratios as a strategy for solving percent problems. The following is the ELL Support activity for this lesson.

ELL SUPPORT

PROBLEM SOLVING

SMALL-GROUP ACTIVITY
5–15 Min

▶ **Summarizing Ratio Concepts**

To provide support for language development, use a graphic organizer like the one shown below to summarize various ways to represent a ratio.

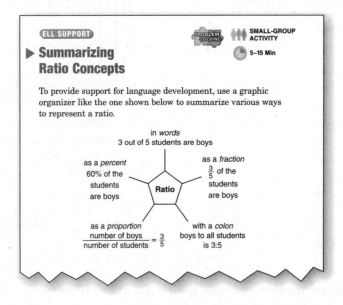

Looking at Grade-Level Goals

Students using *Everyday Mathematics* are expected to master a great deal of mathematical content, but not necessarily the first time the concepts and skills are introduced. The *Everyday Mathematics* curriculum aims for proficiency through repeated exposure over several years of study.

All of the content in *Everyday Mathematics* is important, whether it's being treated for the first time or the fifth time. The *Everyday Mathematics* curriculum is like an intricately woven rug with many threads that appear and reappear to form complex patterns. Students will progress at different rates, so multiple exposures to important content are critical for accommodating individual differences. The program is created to be consistent with how students actually learn mathematics, building understanding over time, first through informal exposure and later through more-formal instruction. It is crucial that students have the opportunity to experience all that the curriculum has to offer in every grade.

To understand where concepts and skills are revisited over time, the unit-specific section of this handbook, beginning on page 47, includes charts for looking at the Grade-Level Goals in each unit. These charts will help you see where you are in the development of the goals—whether the Grade-Level Goal is taught, practiced and applied, or not a focus in the lessons of each unit. The excerpt below can help you understand the information these charts provide.

Map of Number and Numeration Goal 1 for Fourth Grade

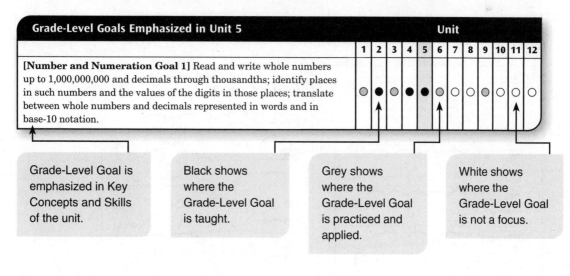

Maintaining Concepts and Skills

The charts discussed in the previous section illustrate where each Grade-Level Goal in a unit is revisited over the course of the year. Sometimes there will be several units in a row that do not address a Grade-Level Goal through either the Key Concepts and Skills emphasized in lessons or through practice. Moreover, as the year progresses, some goals reach the end of their formal development at that grade level. Because students progress at different rates, you may sometimes have students who need to revisit concepts and skills for a particular Grade-Level Goal.

At the end of each unit overview in the unit-specific section of this handbook, you will find a list of *Maintaining Concepts and Skills* activities that you can use to provide students with additional opportunities to explore, review, or practice content. Frequently the list will include references to program routines. These routines, which are revisited throughout the curriculum across the grades, provide a comfortable and convenient way to reinforce, maintain, or further develop concepts and skills for individual students. Blank masters for these routines are included in this handbook beginning on page 133. Examples of helpful strategies are described here.

Frames and Arrows

This routine, which is emphasized in Grades 1 through 3, provides opportunities for students to practice basic and extended addition, subtraction, and multiplication facts. The problems also require students to use algebraic thinking involving patterns, functions, and sequences.

Use these masters to create pages to meet the needs of individual students, or have students create their own problems for classmates to solve.

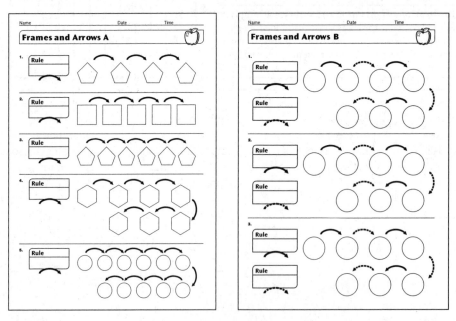

Students use page 142, to practice solving Frames-and-Arrows problems with one rule, and page 143, to practice solving Frames-and-Arrows problems with two rules.

"What's My Rule?"

This routine, which is introduced in Grade 1 and continues through Grade 6, provides opportunities for students to practice basic computation skills and solve problems involving functions. In the upper grades, the functions can be represented visually in graphs and algebraically using variables. In addition to solving teacher-generated problems, students can generate problems for one another to solve. There is a blank master for "What's My Rule?" on page 144 of this handbook.

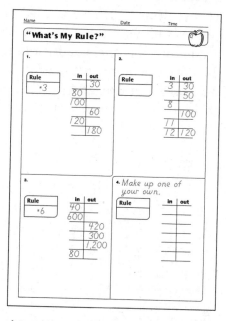

A teacher-generated set of problems that focuses on basic and extended multiplication facts

Name-Collection Boxes

This routine, which is used in Grades 1 through 6, provides the opportunity for students to practice basic computation skills, generate equivalent names for numbers, use grouping symbols, and apply order of operations to numerical expressions. In addition to solving teacher-generated problems, students can generate problems for one another to solve. There is a blank master for name-collection boxes on page 145 of this handbook.

A teacher-generated page that illustrates the variety of ways in which name-collection box problems can be formatted

Number Grids

This routine, which is included in Kindergarten through Grade 3, provides the opportunity for students to explore number relationships and number patterns. Students apply their understanding of these patterns and relationships when they use the number grid as a tool for solving computation problems and when they solve number-grid puzzles. When students become familiar with number-grid puzzles, the puzzles can be extended to include any range of numbers.

Name _____ Date _____ Time _____

Number Grid

-9	-8	-7	-6	-5	-4	-3	-2	-1	0
1	2	3	4	5	6	7	8	9	10
11	12	13	14	15	16	17	18	19	20
21	22	23	24	25	26	27	28	29	30
31	32	33	34	35	36	37	38	39	40
41	42	43	44	45	46	47	48	49	50
51	52	53	54	55	56	57	58	59	60
61	62	63	64	65	66	67	68	69	70
71	72	73	74	75	76	77	78	79	80
81	82	83	84	85	86	87	88	89	90
91	92	93	94	95	96	97	98	99	100
101	102	103	104	105	106	107	108	109	110

-9	-8	-7	-6	-5	-4	-3	-2	-1	0
1	2	3	4	5	6	7	8	9	10
11	12	13	14	15	16	17	18	19	20
21	22	23	24	25	26	27	28	29	30
31	32	33	34	35	36	37	38	39	40
41	42	43	44	45	46	47	48	49	50
51	52	53	54	55	56	57	58	59	60
61	62	63	64	65	66	67	68	69	70
71	72	73	74	75	76	77	78	79	80
81	82	83	84	85	86	87	88	89	90
91	92	93	94	95	96	97	98	99	100
101	102	103	104	105	106	107	108	109	110

Differentiation Masters, page 146

66	67	
	77	78
	87	
96	97	
	107	108

A teacher-generated number-grid puzzle that starts with 66 and 87

991	992	
	1,002	1,003
1,011		1,013
1,021		1,023
1,031		1,033

A teacher-generated number-grid puzzle that starts with 1,002 and 1,033

Projects

This section offers suggestions for how to differentiate the Grade 4 Projects for your students. For each project, you will find three differentiation options: Adjusting the Activity Ideas, ELL Support, and a Writing/Reasoning prompt.

Contents

Projects

Making a Cutaway Globe

Objective To reinforce work with latitude and longitude.

Project 1, along with all other Grade 4 projects, is located at the back of both Volume 1 and Volume 2 of the Grade 4 *Teacher's Lesson Guide*. Use this project during or after Unit 6.

Adjusting the Activity Ideas

◆ Make a Venn diagram comparing and contrasting lines of latitude and longitude.

◆ Have students research the language of global positioning: *degrees, minutes,* and *seconds*.

◆ Have students find the latitude and longitude of key landmarks that they encountered in their World Tour.

ELL Support

Have students use the Math Word Bank template found on page 140 in this handbook. Ask students to write the terms *equator* and *prime meridian*, draw pictures relating to each term, and write other related words. See page 32 of this handbook for more information.

Writing/Reasoning

Have students write a response to the following: Explain the differences and similarities between latitude and longitude lines. (*Sample answer: I know that both latitude and longitude lines are used for global positioning, are measured in degrees, and are imaginary, curved lines. Latitude lines are different because they are complete circles, parallel to each other, described by North and South, and measured from 0 to 90 degrees. Longitude lines are semicircles that begin and end at the poles, described as East and West, and measured from 0 to 180 degrees.*)

Using a Magnetic Compass

Project 2, along with all other Grade 4 projects, is located at the back of both Volume 1 and Volume 2 of the Grade 4 *Teacher's Lesson Guide*. Use this project during or after Unit 6.

Adjusting the Activity Ideas

◆ After north is located, hang signs within the classroom labeling north, east, south, and west.

◆ Review the terms *clockwise* and *counterclockwise.* Make sure that students know which way to rotate for a clockwise turn.

◆ Have students research the difference between the North Pole's position and magnetic north.

ELL Support

◆ Have students use the Math Word Bank template found on page 140 in this handbook. Ask students to write the term *compass*, draw pictures relating to the term, and write other related words. See page 32 of this handbook for more information.

◆ You may want to draw a large compass rose and label the cardinal and intermediate directions. If students know the words for the directions in other languages, add these to the drawing.

Writing/Reasoning

Have students write a response the following: If you were lost or looking for something, how could a magnetic compass be a useful tool? (*Sample answer: A magnetic compass could help me find the direction of things based on where north is.*)

3 A Carnival Game

Objectives To provide opportunities to analyze a cube-tossing game; and to invent a profitable variation.

Project 3, along with all other Grade 4 projects, is located at the back of both Volume 1 and Volume 2 of the Grade 4 *Teacher's Lesson Guide*. Use this project during or after Unit 7.

Adjusting the Activity Ideas

◆ Have students pay with coins for each toss and collect coins as they toss winning cubes onto the class "quilt" mat to help them keep track of their overall gains or losses.

◆ Have students describe how they could make the carnival game a fair game, that is, a game in which players would have the same chance of winning or losing the game.

◆ Have students explain why they might not earn exactly what they predict they will earn.

ELL Support

Have students use the Math Word Bank template found on page 140 in this handbook. Ask students to write the terms *likely, unlikely, certain,* and *impossible,* draw pictures relating to each term, and write other related words. See page 32 of this handbook for more information.

Writing/Reasoning

Have students write a response to the following: What would you say if Ronald said that he landed on yellow 32 times out of 100 tosses while practicing at home? (*Sample answer: I would say that Ronald's results are very unlikely because I would expect the cube to land on yellow 1 time out of every 100 tosses. He might have carefully placed his counter instead of dropping it or he might have colored more than one square yellow.*)

Making a Quilt

Objective To guide students as they explore and apply ideas of pattern, symmetry, rotation, and reflection in the context of quilts.

Project 4, along with all other Grade 4 projects, is located at the back of both Volume 1 and Volume 2 of the Grade 4 *Teacher's Lesson Guide*. Use this project during or after Unit 10.

Adjusting the Activity Ideas

◆ Have students use a transparent mirror to create reflections for patch squares. Then have them rotate the squares in a 9-patch design.

◆ Have students design three different patterns: one with only translations, one with only reflections, and the third with only rotations.

◆ Interview someone who belongs to a quilting group or who has made a quilt. Ask them to describe how they create a design and decide on what colors to use.

ELL Support

Have students collect pictures of quilts or quilt patterns and make a Quilt Museum. Have them describe characteristics that some of the quilts have in common and some of the differences between the quilts. Have English language learners that speak the same language take a minute to discuss the museum in their own language first and then share what they are able to in English.

Writing/Reasoning

Have students write a response to the following: Why do you think color is an important element in quilt making? (*Sample answer: I think color is important because it can change the number of lines of symmetry in an object. Color changes the look of a pattern.*)

Which Soft Drink Is the Best Buy?

Objective To guide students as they calculate the unit price of various soft drinks and decide which is the best buy.

Project 5, along with all other Grade 4 projects, is located at the back of both Volume 1 and Volume 2 of the Grade 4 *Teacher's Lesson Guide*. Use this project during or after Unit 12.

Adjusting the Activity Ideas

◆ Have students use a function table to help determine unit cost.

◆ Collect 10–15 various-shape paper cups and ask students to arrange them by capacity. Then have students check by filling the cups with water or a dry substance that can be poured, such as rice.

◆ Have students compare the cost, nutrients, and calories of a similar amount of regular soda and a more wholesome beverage such as skim milk, fruit juice (not fruit drink), or a yogurt smoothie.

ELL Support

Have students make a large poster of the Big G gallon-equivalency diagram. Have them determine the liter equivalency of each unit in the diagram. If some students are more familiar with metric units, discuss the advantages and disadvantages of the metric system versus the U.S. customary system of measure.

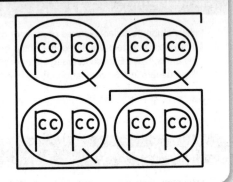

Writing/Reasoning

Have students write a response to the following: How will this project affect how you purchase soft drinks in the future? (*Sample answer: I would definitely find the unit price so I could get the best deal.*)

Project 6

Building and Viewing Structures

 Objectives To provide practice building structures with cubes, given "blueprints" or side views of the structures; and to provide practice representing structures with diagrams.

Project 6, along with all other Grade 4 projects, is located at the back of both Volume 1 and Volume 2 of the Grade 4 *Teacher's Lesson Guide*. Use this project during or after Unit 11.

Adjusting the Activity Ideas

◆ Copy just the Blueprint Mat onto quarter-sheets of paper so that students have a separate building area that is easy to rotate and look at when they are checking their views.

◆ Have students work together in teams to complete the blueprints.

◆ Have students discuss why it might be possible to build two different structures from the same four views, but why it is impossible to build two different structures from the same blueprint. Have them record one set of four views and at least two blueprints that match all four views.

ELL Support

Have students use the Math Word Bank template found on page 140 in this handbook. Ask students to write the term *blueprint*, draw pictures relating to the term, and write other related words. See page 32 of this handbook for more information.

Writing/Reasoning

Have students write a response to the following: Would a top and bottom view look the same for a structure built on a blueprint? Explain. (*Sample answer: Yes; it would look the same, because with a top and bottom view you would be able to see which squares have cubes but not how many cubes are on each square.*)

Project

7 Numbers, Mayan Style

Objectives To introduce the Mayan numeration system; and to provide practice converting between Mayan numerals and base-ten numerals.

Project 7, along with all other Grade 4 projects, is located at the back of both Volume 1 and Volume 2 of the Grade 4 *Teacher's Lesson Guide*. Use this project during or after Unit 5.

Adjusting the Activity Ideas

◆ Have students write an addition problem using Mayan numerals. Have them try to solve it without rewriting the numbers as base-10 numerals.

◆ Have students explain how base-10 blocks would be different and how they might remain the same if they were built to represent Mayan numerals.

◆ Have students research ancient number systems on the Internet. Have them choose one system to highlight on a poster for the classroom.

ELL Support

Have students use the Math Word Bank template found on page 140 in this handbook. Ask students to write the term *place value,* draw pictures relating to the term, and write other related words. See page 32 of this handbook for more information.

Writing/Reasoning

Have students write a response to the following: What are some of the similarities and differences between the Mayan number system and our base-ten numeration system? (*Sample answers: Our number system has ten symbols, and the Mayan number system has only two symbols. We read Mayan numbers from top to bottom, and we read our numbers from left to right. In both systems, the symbols represent numbers and where the symbols are placed tells you what the number is. Each place in our system is 10 times the place to its right. Each place in the Mayan system is 20 times the place below it.*)

Algorithm

1

Project

U.S. Traditional Addition

◎ **Objective** To introduce U.S. traditional addition.

Algorithm Project 1, U.S. Traditional Addition, along with all other Grade 4 projects, is located at the back of both Volume 1 and Volume 2 of the Grade 4 *Teacher's Lesson Guide.* Use this project after Lesson 2-7.

Adjusting the Activity Ideas

◆ Have students start with a 2-digit example to allow them to focus on the procedure.

◆ Have students use a place-value chart when using this method to keep track of the values of the digits.

◆ Have students use base-10 blocks to illustrate each step of U.S. traditional addition. Use base-10 blocks either before or after recording the steps on paper.

ELL Support

Have students mark the appropriate columns with the place-value terms ones, tens, and hundreds. This will help them to remember the values of the digits they are adding.

Writing/Reasoning

Have students write a response to the following: How would you explain this way of adding to someone else? (*Sample answer: I would tell them that you need to work from right to left. Imagine you are adding two 2-digit numbers. First add the ones column. If the answer is only one digit, you write it underneath the line. If the answer is two digits, you write the ones digit underneath the line and the tens digit above the tens column because this is really a ten. Then add the digits in the tens column. If there is a digit above, you need to add that in. Write the ones digit in the tens column underneath because this is really a ten and if there is a tens digit, write it in the hundreds column underneath because this is really a hundred. The number below the line is your answer.*)

Algorithm

2

Project

U.S. Traditional Addition: Decimals

Objective To introduce U.S. traditional addition with decimals.

Algorithm Project 2, U.S. Traditional Addition: Decimals, along with all other Grade 4 projects, is located at the back of both Volume 1 and Volume 2 of the Grade 4 *Teacher's Lesson Guide.* Use this project after Lesson 4-5.

Adjusting the Activity Ideas

◆ Have students start with a 2-digit example to allow them to focus on the procedure.

◆ Have students use a place-value chart when using this method to keep track of the values of the digits and line up the decimal points.

◆ Have students use base-10 blocks to illustrate each step of U.S. traditional addition. Use base-10 blocks either before or after recording the steps on paper.

ELL Support

Have students mark the appropriate columns with the place-value terms ones, tens, hundreds, and, to the right of the decimal point, tenths and hundredths. This will help them to remember the values of the digits they are adding.

Writing/Reasoning

Have students write a response to the following: How is adding with decimals different from adding with whole numbers? (*Sample answer: Adding with decimals is different from adding with whole numbers because you need to line up the decimal points before you start adding. After you finish adding you need to put the decimal point in the answer by lining it up with the other decimal points.*)

Algorithm Project 3, U.S. Traditional Subtraction, along with all other Grade 4 projects, is located at the back of both Volume 1 and Volume 2 of the Grade 4 *Teacher's Lesson Guide*. Use this project after Lesson 2-9.

Adjusting the Activity Ideas

◆ Have students start with a 2-digit example to allow them to focus on the procedure.

◆ Have students use a place-value chart when using this method to keep track of the values of the digits.

◆ Have students use base-10 blocks to illustrate each step of U.S. traditional subtraction. Use base-10 blocks either before or after recording the steps on paper.

ELL Support

Have students mark the appropriate columns with the place-value terms ones, tens, and hundreds. This will help them to remember the values of the digits they are subtracting.

Writing/Reasoning

Have students write a response to the following: How would you explain this way of subtracting to someone else? (*Sample answer: I would tell them that you need to work from right to left. Instead of doing all the trades beforehand, you check if you need to regroup as you go and subtract one column at a time.*)

Algorithm

4

Project

U.S. Traditional Subtraction: Decimals

Objective To introduce U.S. traditional subtraction with decimals.

Algorithm Project 4, U.S. Traditional Subtraction: Decimals, along with all other Grade 4 projects, is located at the back of both Volume 1 and Volume 2 of the Grade 4 *Teacher's Lesson Guide*. Use this project after Lesson 4-6.

Adjusting the Activity Ideas

◆ Have students start with a 2-digit example to allow them to focus on the procedure.

◆ Have students use a place-value chart when using this method to keep track of the values of the digits and line up the decimal points.

◆ Have students use base-10 blocks to illustrate each step of U.S. traditional subtraction. Use base-10 blocks either before or after recording the steps on paper.

ELL Support

Have students mark the appropriate columns with the place-value terms ones, tens, hundreds, and, to the right of the decimal point, tenths and hundredths. This will help them to remember the values of the digits they are subtracting.

Writing/Reasoning

Have students write a response to the following: How would you explain this way of subtracting to someone else? (*Sample answer: I would tell them that you need to work from right to left and make sure the decimal points are lined up. Instead of doing all the trades beforehand, you check if you need to regroup as you go and subtract one column at a time.*)

Algorithm

5

Project

U.S. Traditional Multiplication

◎ **Objective** To introduce U.S. traditional multiplication.

Algorithm Project 5, U.S. Traditional Multiplication, along with all other Grade 4 projects, is located at the back of both Volume 1 and Volume 2 of the Grade 4 *Teacher's Lesson Guide*. Use this project after Lesson 5-7.

Adjusting the Activity Ideas

◆ Have students use a place-value chart when using this method to keep track of the values of the digits.

◆ Have students use a multiplication/division facts table and their knowledge of fact extensions to help them solve problems using U.S. traditional multiplication.

◆ Have students model each step of U.S. traditional multiplication by using the partial-products method simultaneously.

ELL Support

Have students mark the appropriate columns with the place-value terms ones, tens, hundreds, and thousands. This will help them to remember the values of the digits they are multiplying.

Writing/Reasoning

Have students write a response to the following: How would you explain this way of multiplying to someone else? (*Sample answer: It is similar to partial-products except you multiply from right to left. Instead of writing each product on a separate line, you write each digit of the partial products in the appropriate place-value column.*)

Algorithm

6

Project

U.S. Traditional
Multiplication: Decimals

Objective To introduce U.S. traditional multiplication for
decimals in a money context.

Algorithm Project 6, U.S. Traditional Multiplication: Decimals, along with all
other Grade 4 projects, is located at the back of both Volume 1 and Volume 2 of
the Grade 4 *Teacher's Lesson Guide*. Use this project after Lesson 9-8.

Adjusting the Activity Ideas

◆ Have students use a place-value chart when using this method to keep track of the values of the digits.

◆ Have students model each step of U.S. traditional multiplication by using the money method simultaneously.

◆ Have students ignore the decimal points and solve the problem. Then they can use estimation to place the decimal point in the answer.

ELL Support

To support using estimation to place the decimal point in a product, have students consider a problem such as $2.9 * 42$. Guide students to conclude that 2.9 is about 3, and 42 is about 40, so the answer should be about $3 * 40 = 120$. When the decimal point is placed in 1218 (the product obtained when the decimal points are ignored), students conclude that $2.9 * 42 = 121.8$. Encourage students to describe their thinking aloud as they estimate and place the decimal point in the product.

Writing/Reasoning

Have students write a response to the following: How is multiplying with decimals different from multiplying with whole numbers? (*Sample answer: Multiplying with decimals is different from multiplying with whole numbers because you need to add in the decimal point after you get your answer. You can count the total number of places to the right of the decimal point in each number you multiplied and then count over that many places in the answer to place the decimal point.*)

Algorithm

7

Project

U.S. Traditional Long Division, Part 1

◎ **Objective** To introduce U.S. traditional long division.

Algorithm Project 7, U.S. Traditional Long Division, Part 1, along with all other Grade 4 projects, is located at the back of both Volume 1 and Volume 2 of the Grade 4 *Teacher's Lesson Guide*. Use this project after Lesson 6-10.

Adjusting the Activity Ideas

◆ Encourage students to think of these problems in terms of fair sharing. Allow them to use manipulatives, like pencils in boxes or money, to illustrate this method.

◆ Have students use a place-value chart when using this method to keep track of the values of the digits.

◆ Have students use base-10 blocks to illustrate each step of U.S. traditional division. Use base-10 blocks either before or after recording the steps on paper.

ELL Support

Have students mark the appropriate columns with the place-value terms ones, tens, and hundreds. This will help them to remember the values of the digits they are dividing.

Writing/Reasoning

Have students write a response to the following: How would you explain this way of dividing to someone else? (*Sample answer: It is similar to partial quotients but instead of writing each partial quotient to the right of the problem, you write the quotients above the appropriate digit in the dividend. You still write each product underneath the dividend and subtract. You might have to regroup. You continue dividing until you cannot divide anymore.*)

Algorithm

8

Project

U.S. Traditional Long Division, Part 2

Objective To extend U.S. traditional long division with single-digit divisors to four- and five-digit dividends and dividends in dollars-and-cents notation.

Algorithm Project 8, U.S. Traditional Long Division, Part 2, along with all other Grade 4 projects, is located at the back of both Volume 1 and Volume 2 of the Grade 4 *Teacher's Lesson Guide*. Use this project after Lesson 9-9 and Algorithm Project 7.

Adjusting the Activity Ideas

◆ Have students use a place-value chart when using this method to keep track of the values of the digits.

◆ Have students model each step of U.S. traditional division by using the money method simultaneously.

◆ Have students ignore the decimal points and solve the problem. Then they can use estimation to place the decimal point in the answer.

ELL Support

Have students mark the appropriate columns with the place-value terms ones, tens, hundreds, and, to the right of the decimal point, tenths and hundredths. This will help them to remember the values of the digits they are dividing.

Writing/Reasoning

Have students write a response to the following: How is dividing with decimals different from dividing with whole numbers? (*Sample answer: Dividing with decimals is different from dividing with whole numbers because you need to place the decimal point after you get your answer. You can place the decimal point directly above the decimal point in the dividend.*)

Activities and Ideas for Differentiation

This section highlights Part 1 activities that support differentiation, optional Part 3 Readiness, Enrichment, Extra Practice, and ELL Support activities built into the lessons of the Grade 4 *Teacher's Lesson Guide,* and specific ideas for vocabulary development and games modifications. Provided in each unit is a chart showing where the Grade-Level Goals emphasized in that unit are addressed throughout the year. Following the chart, there are suggestions for maintaining concepts and skills to ensure that students continue working toward those Grade-Level Goals.

Contents

In this unit, students describe, compare, and classify plane figures. This section summarizes opportunities for supporting multiple learning styles and ability levels. Use these suggestions to develop a differentiation plan for Unit 1.

Part 1 Activities That Support Differentiation

Below are examples of Unit 1 activities that highlight some of the general instructional strategies that are hallmarks of a differentiated classroom. These strategies will help you support, emphasize, and enhance lesson content to make sure all your students are engaged in the mathematics at the highest possible level. For more information about general differentiation strategies that accommodate the diverse needs of today's classrooms, see the essay on pages 8–16 of this handbook.

Lesson	Activity	Strategy
1◆1	As each section of the *Student Reference Book* is discussed, write its name on the board.	Recording key ideas
1◆2	Students look for objects in the classroom that model geometric shapes.	Making connections to everyday life
1◆3	Students build geometric figures with straws and twist-ties.	Modeling concretely
1◆4	Associate parallel lines with examples in everyday life, such as line segments on notebook paper or the rails of train tracks.	Making connections to everyday life
1◆6	Students inscribe a square in a circle using a compass and straightedge.	Modeling physically
1◆8	Students copy a line segment and construct an inscribed, regular hexagon using a compass and straightedge.	Modeling physically

Vocabulary Development

The list below identifies the Key Vocabulary terms from this unit. The lesson in which each term is defined is indicated next to the term. Some of these terms or their homophones are used outside of mathematics. Consider adding other words as appropriate for developing understanding of the context of the lessons.

Lessons include suggestions for helping English language learners understand and develop vocabulary. For more information, see pages 17–19 of this handbook.

Key Vocabulary

angle 1◆3	*kite 1◆3	polygon 1◆5
center (of a circle) 1◆6	line 1◆2	quadrangle 1◆3
circle 1◆6	line segment 1◆2	quadrilateral 1◆3
*compass 1◆6	*n*-gon 1◆5	radius 1◆7
concentric circles 1◆7	nonagon 1◆5	*ray 1◆2
congruent 1◆7	nonconvex or concave 1◆5	rectangle 1◆3
convex 1◆5	octagon 1◆5	regular polygon (*regular) 1◆5
endpoint 1◆2	parallel line segments 1◆4	rhombus 1◆3
equilateral triangle 1◆5	parallel lines 1◆4	right angle (*†right) 1◆3
heptagon 1◆5	parallel rays 1◆4	*†side 1◆5
hexagon 1◆5	parallelogram 1◆3	square 1◆3
inscribed square 1◆6	pentagon 1◆5	trapezoid 1◆3
interior 1◆5	perpendicular line segments 1◆4	triangle 1◆3
intersect 1◆4	*point 1◆2	vertex (vertices) 1◆3

* Discuss the everyday and mathematical meanings of the words that are marked with an asterisk.

† For words marked with a dagger, write the words and their homophones on the board. For example, *right, write, rite* and *wright;* and *side* and *sighed*. Discuss and clarify the meaning of each.

◆ As each word is introduced in the lesson, write the word on the board and discuss its meaning.

◆ List the words on a Math Word Wall for students to see. As each word is introduced in the lesson, add a picture next to the word on the Word Wall.

◆ Use the vocabulary words regularly when teaching lessons, and encourage students to use the words in their discussions.

 **Games**

Below are suggested Unit 1 game adaptations. For more information about implementing games in a differentiated classroom, see pages 20–25 of this handbook.

Game: *Addition Top-It* (Extended-Facts Version)

Skill Practiced: **Solve addition fact extensions.** [Operations and Computation Goal 1]

Modification	Purpose of Modification
Students play the original basic-facts version of *Addition Top-It*.	Students develop automaticity with basic addition facts. [Operations and Computation Goal 1]
Players turn over three cards and find the sum on each turn.	Students solve addition problems with three addends. [Operations and Computation Goal 2]

Game: *Subtraction Top-It* (Extended-Facts Version)

Skill Practiced: **Solve subtraction fact extensions.** [Operations and Computation Goal 1]

Modification	Purpose of Modification
Students play the original basic-facts version of *Subtraction Top-It*.	Students develop automaticity with basic subtraction facts. [Operations and Computation Goal 1]
Players attach two 0s to each card that they draw.	Students solve subtraction fact extensions in the 100s. [Operations and Computation Goal 2]

Game: *Polygon Pair-Up*

Skill Practiced: **Describe properties of polygons.** [Geometry Goal 2]

Modification	Purpose of Modification
Players draw one Property Card and use their Geometry Templates to draw a shape with that property. They receive one point if they can draw a shape.	Students draw polygons according to polygon properties. [Geometry Goal 2]
Each partnership needs two sets of Property Cards. On each turn, players draw two Property Cards and one Polygon Card. They can use one or two Property Cards to make a match. At the end, the player who has collected the most cards wins.	Students describe properties of polygons. [Geometry Goal 2]

 **Math Boxes**

Suggestions for using Math Boxes to meet individual needs begin on page 26 of this handbook. There are blank masters for Math Boxes on pages 134–139.

Using Part 3 of the Lessons

Use your professional judgment, along with assessment results, to determine whether the whole class, small groups, or individual students might benefit from these Unit 1 activities. Consider using the Part 3 Planning Master found on page 152 of this handbook to record your plans.

Readiness Activities

Lesson	Activity	Purpose of Activity
1♦1	Play *Top-It*.	Explore the use of the relation symbols <, >, and =. [Number and Numeration Goal 6]
1♦2	Make line segments on a geoboard.	Explore the characteristics of a line segment; highlight the difference between a line and a line segment. [Geometry Goal 1]
1♦3	Sort pattern blocks according to rules.	Explore properties of quadrangles. [Geometry Goal 2]
1♦4	Make parallel line segments on a geoboard.	Gain experience with parallel line segments. [Geometry Goal 1]
1♦5	Make polygons on a geoboard.	Gain experience with polygons and their properties. [Geometry Goal 2]
1♦5	Read *The Greedy Triangle*, and use straws and twist-ties to make the polygons from the story.	Explore classifying polygons by number of sides or angles. [Geometry Goal 2]
1♦8	Identify a regular hexagon in a design.	Explore the concept of regular polygons. [Geometry Goal 2]

English Language Learners Support Activities

Lesson	Activity	Purpose of Activity
1♦2	Generate a list of mathematical *tools*.	Clarify the mathematical and everyday uses of the term; record key ideas on the board. [Geometry Goal 1]
1♦3	Use a Venn diagram to compare and contrast *quadrangles*.	Connect a new term to existing vocabulary; use a graphic organizer to describe the characteristics of quadrangles. [Geometry Goal 2]
1♦7	Add *intersect* to the Math Word Bank.	Make connections between a term and terms students know; use visual models to represent the term. [Geometry Goal 1]

Enrichment Activities

Lesson	Activity	Purpose of Activity
1•1	Search for mathematical symbols in the *Student Reference Book*.	Explore mathematical symbols and their meanings. [Number and Numeration Goal 6]
1•2	Solve a collinear-points puzzle.	Explore characteristics of lines. [Geometry Goal 1]
1•2	Play *Sprouts*.	Explore line segments and points. [Geometry Goal 1]
1•3	Solve embedded-polygon puzzles.	Apply understanding of properties of rectangles, triangles, and squares. [Geometry Goal 2]
1•4	Solve a puzzle by rearranging the sides of a rectangle to make two squares.	Apply understanding of the properties of parallelograms. [Geometry Goal 2]
1•4	Play *Sz'kwa*.	Apply understanding of intersecting line segments. [Geometry Goal 1]
1•5	Compare, contrast, and describe the properties of kites and rhombuses.	Apply understanding of properties of kites and rhombuses. [Geometry Goal 2]
1•6	Draw inscribed squares.	Apply understanding of inscribed polygons. [Geometry Goal 2]
1•7	Construct tangent circles using a compass.	Apply understanding of circle constructions. [Geometry Goal 2]
1•7	Inscribe polygons in circles.	Apply understanding of polygon properties. [Geometry Goal 2]
1•8	Use a compass to create hexagram designs.	Apply understanding of hexagons inscribed in circles. [Geometry Goal 2]

Extra Practice Activities

Lesson	Activity	Purpose of Activity
1•3	Solve *5-Minute Math* problems involving characteristics of 2-dimensional shapes.	Practice using vocabulary for properties of 2-dimensional shapes. [Geometry Goal 2]
1•6	Create circle designs with a compass based on *Ed Emberley's Picture Pie: A Cut and Paste Drawing Book*.	Practice using a compass to draw circles. [Geometry Goal 2]
1•8	Inscribe an equilateral triangle in a circle.	Practice inscribing polygons in circles. [Geometry Goal 2]

Looking at Grade-Level Goals

Everyday Mathematics develops concepts and skills over time. Below is a chart showing where the Grade-Level Goals emphasized in this unit are addressed throughout the year. Use the chart to help you determine which Maintaining Concepts and Skills activities on page 54 to utilize to ensure that students continue working toward these Grade-Level Goals.

- ● Grade-Level Goal is taught.
- ◐ Grade-Level Goal is practiced and applied.
- ○ Grade-Level Goal is not a focus.

Grade-Level Goals Emphasized in Unit 1

	Unit											
	1	2	3	4	5	6	7	8	9	10	11	12
[Measurement and Reference Frames Goal 1] Estimate length with and without tools; measure length to the nearest $\frac{1}{4}$ inch and $\frac{1}{2}$ centimeter; use tools to measure and draw angles; estimate the size of angles without tools.	●	◐	○	●	○	●	○	●	○	○	○	○
[Geometry Goal 1] Identify, draw, and describe points, intersecting and parallel line segments and lines, rays, and right, acute, and obtuse angles.	●	◐	◐	◐	○	●	●	◐	◐	◐	○	○
[Geometry Goal 2] Describe, compare, and classify plane and solid figures, including polygons, circles, spheres, cylinders, rectangular prisms, cones, cubes, and pyramids, using appropriate geometric terms including *vertex, base, face, edge,* and *congruent.*	●	◐	◐	◐	○	●	●	●	●	●	●	○

Maintaining Concepts and Skills

All of the goals addressed in this unit will be addressed again in later units. Here are several suggestions for maintaining concepts and skills until they are formally revisited.

Measurement and Reference Frames Goal 1

◆ Have students make geometric constructions using their compasses to measure side lengths and angle openings. See the *Student Reference Book* geometry section for more information about constructions.

Geometry Goal 1

◆ Have students play *Sprouts* or *Sz'kwa*.

◆ Have students make line segments and parallel line segments on a geoboard. See the Readiness activities in Lessons 1-2 and 1-4 for more information.

Geometry Goal 2

◆ Have students play *Polygon Pair-Up*.

◆ Read *The Greedy Triangle* and have students use straws and twist-ties to make the polygons from the story. See the Readiness activity in Lesson 1-5 for more information.

◆ Have students build polygons with straws and twist-ties and describe the properties of the polygons they build.

Assessment

See page 54 in the *Assessment Handbook* for modifications to the written portion of the Unit 1 Progress Check.

Additionally, see pages 55–59 for modifications to the open-response task and selected student work samples.

Unit 2 Activities and Ideas for Differentiation

In this unit, students review place-value concepts and computation skills. They also expand their experiences with data collection, organization, display, and analysis. This section summarizes opportunities for supporting multiple learning styles and ability levels. Use these suggestions to develop a differentiation plan for Unit 2.

Part 1 Activities That Support Differentiation

Below are examples of Unit 2 activities that highlight some of the general instructional strategies that are hallmarks of a differentiated classroom. These strategies will help you support, emphasize, and enhance lesson content to make sure all of your students are engaged in the mathematics at the highest possible level. For more information about general differentiation strategies that accommodate the diverse needs of today's classrooms, see the essay on pages 8–16 of this handbook.

Lesson	Activity	Strategy
2◆1	Summarize the five categories of number uses and record them on the board.	Recording key ideas
2◆3	Students use place-value charts to explore place value.	Using organizational tools
2◆5	Students predict the number of raisins in a small box and collect data to check predictions.	Modeling concretely
2◆6	Students make a line plot with stick-on notes.	Modeling concretely
2◆7	Students use the partial-sums and column-addition algorithms to solve addition problems.	Incorporating and validating a variety of methods
2◆9	Students use the trade-first and partial-differences algorithms to solve subtraction problems.	Incorporating and validating a variety of methods

Vocabulary Development

The list below identifies the Key Vocabulary terms from this unit. The lesson in which each term is defined is indicated next to the term. Some of these terms or their homophones are used outside of mathematics. Consider adding other words as appropriate for developing understanding of the context of the lessons.

Lessons include suggestions for helping English language learners understand and develop vocabulary. For more information, see pages 17–19 of this handbook.

Key Vocabulary

ballpark estimate **2◆7**	guess **2◆5**	partial-differences method **2◆9**
bar graph (*bar) **2◆8**	*landmark **2◆5**	partial-sums method (†sum) **2◆7**
column-addition method **2◆7**	line plot **2◆6**	*†place **2◆3**
counting number **2◆3**	maximum **2◆5**	*range **2◆5**
*digit **2◆3**	*median **2◆6**	tally chart **2◆5**
equivalent name **2◆2**	minimum **2◆5**	trade-first method **2◆9**
estimate **2◆5**	†mode **2◆5**	whole number (*†whole) **2◆3**
expanded notation **2◆3**	name-collection box **2◆2**	

* Discuss the everyday and mathematical meanings of the words that are marked with an asterisk.

† For words marked with a dagger, write the words and their homophones on the board. For example, *mode* and *mowed; sum* and *some; place* and *plaice;* and *whole* and *hole.* Discuss and clarify the meaning of each.

◆ As each word is introduced in the lesson, write the word on the board and discuss its meaning.

◆ List the words on a Math Word Wall for students to see. As each word is introduced in the lesson, add a picture next to the word on the Word Wall.

◆ Use the vocabulary words regularly when teaching lessons, and encourage students to use the words in their discussions.

 **Games**

Below are suggested Unit 2 game adaptations. For more information about implementing games in a differentiated classroom, see pages 20–25 of this handbook.

Game: *Name That Number*

Skill Practiced: Write equivalent names for numbers. [Number and Numeration Goal 4]

Modification	Purpose of Modification
Players draw six cards instead of five to increase their options.	Students write equivalent names for numbers. [Number and Numeration Goal 4]
Players represent the target number using both addition and subtraction in each solution.	Students solve addition and subtraction problems. [Operations and Computation Goal 2]

Game: *High-Number Toss*

Skill Practiced: Compare numbers. [Number and Numeration Goal 6]

Modification	Purpose of Modification
Players use all four spaces as digits so that each number they build is in the thousands.	Students read and write whole numbers. [Number and Numeration Goal 1]
After all of the spaces are filled in, players can make one "switch," trading the places for two of the digits. Then they record and compare their numbers.	Students compare numbers. [Number and Numeration Goal 6]

Game: *Subtraction Target Practice*

Skill Practiced: Solve subtraction problems. [Number and Numeration Goal 1; Operations and Computation Goal 2]

Modification	Purpose of Modification
Players turn over only one card on each turn. They subtract that number or that many tens from their total. For example, a player turns over a 3 and can subtract either 3 or 30.	Students solve subtraction problems. [Number and Numeration Goal 1; Operations and Computation Goal 2]
Players draw two cards and either subtract a 2-digit number that they make with the cards or subtract the sum of the two cards from their total.	Students solve addition and subtraction problems. [Number and Numeration Goal 1; Operations and Computation Goal 2]

 **Math Boxes**

Suggestions for using Math Boxes to meet individual needs begin on page 26 of this handbook. There are blank masters for Math Boxes on pages 134–139.

Using Part 3 of the Lessons

Use your professional judgment, along with assessment results, to determine whether the whole class, small groups, or individual students might benefit from these Unit 2 activities. Consider using the Part 3 Planning Master found on page 152 of this handbook to record your plans.

Readiness Activities

Lesson	Activity	Purpose of Activity
2♦1	Solve Frames-and-Arrows problems.	Use rules to explore linear intervals. [Patterns, Functions, and Algebra Goal 1]
2♦2	Sort dominoes by sums.	Explore equivalent names for numbers. [Number and Numeration Goal 4]
2♦4	Use a Compact Place-Value Flip Book to solve problems.	Gain experience identifying the place value of digits in large numbers. [Number and Numeration Goal 1]
2♦5	Use tally marks to record dice-roll data.	Gain experience with tally marks. [Data and Chance Goal 1]
2♦6	Order number cards and find the middle value.	Explore data landmarks. [Data and Chance Goal 2]
2♦7	Solve parts-and-total number stories using base-10 blocks.	Explore parts-and-total number stories. [Operations and Computation Goal 2]
2♦8	Construct a bar graph using pattern blocks.	Explore bar graphs. [Data and Chance Goal 1]
2♦9	Subtract by counting up on a number line.	Explore a subtraction algorithm. [Operations and Computation Goal 2]

English Language Learners Support Activities

Lesson	Activity	Purpose of Activity
2♦1	Explore a world map to introduce the *World Tour*.	Use the map as a visual reference for new terms. [Measurement and Reference Frames Goal 4]
2♦3	Add *counting numbers* and *whole numbers* to the Math Word Bank.	Make connections between new terms and terms students know; use visual models to represent these terms. [Number and Numeration Goal 1]
2♦6	Add *median* to the Math Word Bank.	Make connections between a new term and terms students know; use visual models to represent the term. [Data and Chance Goal 2]
2♦7	Add *ballpark estimate* to the Math Word Bank.	Make connections between a new term and terms students know; use visual models to represent these terms. [Operations and Computation Goal 6]

Enrichment Activities

Lesson	Activity	Purpose of Activity
2•1	Find missing numbers on a number line.	Apply understanding of intervals. [Patterns, Functions, and Algebra Goal 1]
2•2	Solve pan-balance problems.	Apply understanding of equivalent names for numbers. [Number and Numeration Goal 4]
2•3	Solve number-grid puzzles.	Apply understanding of the base-ten place-value system. [Number and Numeration Goal 1]
2•4	Decipher a place-value code in another base.	Apply understanding of place value. [Number and Numeration Goal 1]
2•5	Make a prediction based on a sample of data.	Apply understanding of data landmarks. [Data and Chance Goal 2]
2•6	Compare family-size data among classes.	Explore organizing and summarizing data. [Data and Chance Goal 2]
2•7	Write addition number stories.	Apply understanding of addition algorithms. [Operations and Computation Goal 2]
2•8	Analyze head-size data to determine hat sizes.	Explore the application of data. [Data and Chance Goal 2]
2•9	Solve addition and subtraction problems with number tiles.	Apply computation and estimation skills. [Operations and Computation Goal 2]
2•9	Write subtraction number stories.	Apply understanding of subtraction algorithms. [Operations and Computation Goal 2]

Extra Practice Activities

Lesson	Activity	Purpose of Activity
2•1	Solve Frames-and-Arrows problems.	Practice extending numeric patterns. [Patterns, Functions, and Algebra Goal 1]
2•2	Complete name-collection boxes.	Practice writing equivalent names for numbers. [Number and Numeration Goal 4]
2•2	Solve 5-Minute Math problems involving Roman numerals.	Practice writing equivalent names for numbers. [Number and Numeration Goal 4]
2•3	Solve place-value problems.	Practice place-value skills. [Number and Numeration Goal 1]
2•5	Solve 5-Minute Math problems involving landmarks.	Practice identifying data landmarks. [Data and Chance Goal 2]

Looking at Grade-Level Goals

Everyday Mathematics develops concepts and skills over time. Below is a chart showing where the Grade-Level goals emphasized in this unit are addressed throughout the year. Use the chart to help you determine which Maintaining Concepts and Skills activities on page 61 to utilize to ensure that students continue working toward these Grade-Level Goals.

- ● Grade-Level Goal is taught.
- ◐ Grade-Level Goal is practiced and applied.
- ○ Grade-Level Goal is not a focus.

Grade-Level Goals Emphasized in Unit 2	1	2	3	4	5	6	7	8	9	10	11	12
[Number and Numeration Goal 1] Read and write whole numbers up to 1,000,000,000 and decimals through thousandths; identify places in such numbers and the values of the digits in those places; translate between whole numbers and decimals represented in words and in base-10 notation.	◐	●	◐	◐	●	◐	◐	○	◐	◐	◐	○
[Number and Numeration Goal 4] Use numerical expressions involving one or more of the basic four arithmetic operations and grouping symbols to give equivalent names for whole numbers.	◐	●	◐	◐	●	◐	◐	◐	◐	◐	◐	○
[Operations and Computation Goal 1] Demonstrate automaticity with addition and subtraction fact extensions.	◐	●	◐	◐	○	◐	○	◐	◐	◐	◐	◐
[Operations and Computation Goal 2] Use manipulatives, mental arithmetic, paper-and-pencil algorithms and models, and calculators to solve problems involving the addition and subtraction of whole numbers and decimals through hundredths; describe the strategies used and explain how they work.	◐	●	●	●	●	◐	◐	◐	●	◐	◐	◐
[Operations and Computation Goal 6] Make reasonable estimates for whole number and decimal addition and subtraction problems and whole number multiplication and division problems; explain how the estimates were obtained.	○	●	●	●	●	◐	◐	◐	●	◐	◐	◐
[Data and Chance Goal 1] Collect and organize data or use given data to create charts, tables, graphs, and line plots.	○	○	◐	○	◐	◐	◐	◐	◐	◐	◐	◐
[Data and Chance Goal 2] Use the maximum, minimum, range, median, mode, and graphs to ask and answer questions, draw conclusions, and make predictions.	◐	●	◐	◐	●	◐	◐	◐	●	◐	◐	◐
[Patterns, Functions, and Algebra Goal 1] Extend, describe, and create numeric patterns; describe rules for patterns and use them to solve problems; use words and symbols to describe and write rules for functions that involve the four basic arithmetic operations and use those rules to solve problems.	◐	●	●	◐	●	◐	○	◐	◐	◐	◐	◐
[Patterns, Functions, and Algebra Goal 2] Use conventional notation to write expressions and number sentences using the four basic arithmetic operations; determine whether number sentences are true or false; solve open sentences and explain the solutions; write expressions and number sentences to model number stories.	○	●	◐	◐	◐	◐	○	○	◐	◐	◐	○

Maintaining Concepts and Skills

Many of the goals addressed in this unit will be addressed again in later units. Those goals marked with an asterisk (*) are only addressed in future units as practice and application. Here are several suggestions for maintaining concepts and skills until they are formally revisited.

Number and Numeration Goal 1

◆ Have students use a Compact Place-Value Flip Book to solve problems. See the Readiness activity in Lesson 2-4 for more information.

Operations and Computation Goal 1*

◆ Have students play *Addition Top-It* (Extended-Facts Version) and *Subtraction Top-It* (Extended-Facts Version).

Operations and Computation Goal 2

◆ Have students play *Subtraction Target Practice.*

◆ Have students solve parts-and-total number stories using base-10 blocks. See the Readiness activity in Lesson 2-7 for more information.

Patterns, Functions, and Algebra Goal 1

◆ Use Frames-and-Arrows masters A and B on pages 142 and 143 of this handbook to create practice problems.

Assessment

See page 62 in the *Assessment Handbook* for modifications to the written portion of the Unit 2 Progress Check.

Additionally, see pages 63–67 for modifications to the open-response task and selected student work samples.

Activities and Ideas for Differentiation

In this unit, students practice multiplication facts and extend their work with number sentences and variables. This section summarizes opportunities for supporting multiple learning styles and ability levels. Use these suggestions to develop a differentiation plan for Unit 3.

Part 1 Activities That Support Differentiation

Below are examples of Unit 3 activities that highlight some of the general instructional strategies that are hallmarks of a differentiated classroom. These strategies will help you support, emphasize, and enhance lesson content to make sure all of your students are engaged in the mathematics at the highest possible level. For more information about general differentiation strategies that accommodate the diverse needs of today's classrooms, see the essay on pages 8–16 of this handbook.

Lesson	Activity	Strategy
3•1	Students solve problems according to rules and use a table to organize numbers.	Using organizational tools
3•4	Students collect and graph data about their performance on a multiplication-facts test.	Making connections to everyday life
3•5	Students use what they know about multiplication facts to solve division fact problems.	Building on prior knowledge
3•7	Students use a cloth tape measure and the globe scale to find air distances between locations on the globe.	Modeling physically
3•9	Record number sentences on the board with both words and symbols.	Using visual references

Vocabulary Development

The list below identifies the Key Vocabulary terms from this unit. The lesson in which each term is defined is indicated next to the term. Some of these terms or their homophones are used outside of mathematics. Consider adding other words as appropriate for developing understanding of the context of the lessons.

Lessons include suggestions for helping English language learners understand and develop vocabulary. For more information, see pages 17–19 of this handbook.

Key Vocabulary

composite number **3♦2**	multiplication facts **3♦2**	remainder **3♦5**
dividend **3♦5**	number sentence **3♦9**	*rule **3♦1**
†divisor **3♦5**	open sentence (*open) **3♦11**	*solution **3♦11**
fact family **3♦5**	output **3♦1**	solve **3♦11**
factor pair (†pair) **3♦2**	parentheses **3♦10**	square numbers **3♦2**
*factors **3♦2**	percent **3♦3**	true number sentence **3♦9**
false number sentence **3♦9**	prime number (*prime) **3♦2**	turn-around facts **3♦2**
function machine **3♦1**	*products **3♦2**	variable **3♦11**
input **3♦1**	quotient **3♦5**	"What's My Rule?" **3♦1**
multiples **3♦2**		

* Discuss the everyday and mathematical meanings of the words that are marked with an asterisk.

† For words marked with a dagger, write the words and their homophones on the board. For example, *divisor* and *deviser* and *pair, pare,* and *pear.* Discuss and clarify the meaning of each.

◆ As each word is introduced in the lesson, write the word on the board and discuss its meaning.

◆ List the words on a Math Word Wall for students to see. As each word is introduced in the lesson, add a picture next to the word on the Word Wall.

◆ Use the vocabulary words regularly when teaching lessons, and encourage students to use the words in their discussions.

 **Games**

Below are suggested Unit 3 game adaptations. For more information about implementing games in a differentiated classroom, see pages 20–25 of this handbook.

Game: *Multiplication Top-It*

Skill Practiced: Solve multiplication facts. [Operations and Computation Goal 3]

Modification	Purpose of Modification
Players keep one factor constant and turn over a card for the second factor in each round. If a 5 is the constant, players draw one card and multiply that number by 5.	Students solve a group of multiplication facts with one factor in common. [Operations and Computation Goal 3]
Players use cards 1–9. They turn over three cards, form a 2-digit number, and multiply by the number on the third card.	Students solve multidigit multiplication problems. [Operations and Computation Goal 4]

Game: *Division Arrays*

Skill Practiced: Explore relationships between multiplication and division. [Operations and Computation Goal 7]

Modification	Purpose of Modification
Players draw a card and make a rectangular array with that number of counters. Their score is the number of columns or the number of rows, whichever is smaller. For example, for 18, if they make a 1 × 18 array, their score is 1. If they make a 3 × 6 array, their score is 3.	Students build rectangular arrays and find factors of a given number. [Operations and Computation Goal 7]
Players predict whether they will be able to build a rectangular array. They receive a bonus point if their prediction is correct. They receive another bonus point if they can write a fact family for their array.	Students identify factor pairs for a number and build arrays to find missing factors. [Operations and Computation Goal 7]

Game: *High-Number Toss*

Skill Practiced: Identify place value and compare numbers. [Number and Numeration Goals 1 and 6]

Modification	Purpose of Modification
For each round, players use the same digit in the "Number of zeros" space. They roll the die only three times to complete their numbers.	Students read, write, and compare numbers in a specified place value. [Number and Numeration Goals 1 and 6]
In each round, players make two numbers, and their score is the difference between the numbers. The highest score wins the round.	Students read, write, compare, and find the difference between large numbers. [Number and Numeration Goals 1 and 6; Operations and Computation Goal 2]

 **Math Boxes**

Suggestions for using Math Boxes to meet individual needs begin on page 26 of this handbook. There are blank masters for Math Boxes on pages 134–139.

Using Part 3 of the Lessons

Use your professional judgment, along with assessment results, to determine whether the whole class, small groups, or individual students might benefit from these Unit 3 activities. Consider using the Part 3 Planning Master found on page 152 of this handbook to record your plans.

Readiness Activities

Lesson	Activity	Purpose of Activity
3◆1	Determine relationships between the number of shapes and the number of sides for specified shapes.	Explore the relationships between pairs of numbers. [Patterns, Functions, and Algebra Goal 1]
3◆2	Make rectangular arrays with centimeter cubes.	Gain experience with multiplication facts. [Operations and Computation Goal 3]
3◆3	Explore skip-count patterns on the hundreds grid.	Gain experience with multiplication-fact patterns. [Operations and Computation Goal 3]
3◆4	Find the mean of a data set using stacks of cubes.	Explore finding the mean of a data set. [Data and Chance Goal 2]
3◆5	Play *Division Arrays.*	Explore relationships between multiplication and division. [Operations and Computation Goal 7]
3◆6	Listen to *Nine O'Clock Lullaby* and locate places mentioned in the book.	Explore time zones. [Operations and Computation Goal 2; Data and Chance Goal 2]
3◆7	Measure line segments to the nearest $\frac{1}{2}$ inch using a ruler marked in $\frac{1}{4}$ inches.	Gain experience measuring line segments. [Measurement and Reference Frames Goal 1]
3◆8	Solve number stories using situation diagrams.	Explore solving number stories. [Operations and Computation Goal 2]
3◆9	Solve comparison problems using analogies and mnemonic devices.	Gain experience using relation symbols with comparison problems. [Patterns, Functions, and Algebra Goal 2]
3◆11	Use Fact Triangles to write and solve open sentences.	Explore the concept of open sentences. [Patterns, Functions, and Algebra Goal 2]

English Language Learners Support Activities

Lesson	Activity	Purpose of Activity
3◆2	Add *square numbers* to the Math Word Bank.	Make connections between a new term and terms students know; use visual models to represent the term. [Operations and Computation Goal 3]
3◆5	Create a poster using *dividend, divisor, quotient,* and *remainder.*	Use a student-made poster as a visual reference for new terms. [Operations and Computation Goals 3 and 4]
3◆8	Create a poster of words and phrases commonly used in *number stories.*	Use a student-made poster as a visual reference for new terms. [Operations and Computation Goals 2 and 4]
3◆9	Add *number sentence, true,* and *false* to the Math Word Bank.	Make connections between new terms and terms students know; use visual models to represent the terms. [Patterns, Functions, and Algebra Goal 2]

Enrichment Activities

Lesson	Activity	Purpose of Activity
3•1	Investigate a perimeter pattern for rectangles made of square pattern blocks placed side by side.	Apply understanding of functional relationships. [Patterns, Functions, and Algebra Goal 1]
3•3	Solve a combinations problem.	Apply understanding of multiplication. [Operations and Computation Goal 3]
3•4	Compare mean and median using baseball salaries.	Apply understanding of mean and median. [Data and Chance Goal 2]
3•5	Solve a problem involving sharing a pizza and record the solution using fraction notation.	Explore the relationship between fractions and division. [Operations and Computation Goal 3]
3•6	Play *Seega*.	Explore mathematical connections with Egypt. [Patterns, Functions, and Algebra Goal 1]
3•7	Explore whether changing a map scale changes the actual distance represented by the map.	Apply understanding of map scales. [Operations and Computation Goal 7]
3•8	Solve and write a number story based on an essay.	Apply understanding of solving number stories. [Operation and Computation Goal 2]
3•9	Arrange digits in a puzzle grid to make three true number sentences.	Apply understanding of true number sentences. [Patterns, Functions, and Algebra Goal 2]
3•10	Write number models with parentheses to represent situations.	Apply understanding of the use of parentheses in number models. [Patterns, Functions, and Algebra Goal 3]
3•11	Solve open sentences.	Apply understanding of open sentences. [Patterns, Functions, and Algebra Goal 2]

Extra Practice Activities

Lesson	Activity	Purpose of Activity
3•1	Complete "What's My Rule?" tables.	Practice using words and symbols to describe and write rules for functions. [Patterns, Functions, and Algebra Goal 1]
3•2	Play *Buzz* and *Bizz-Buzz*.	Practice identifying multiples of numbers. [Number and Numeration Goal 3]
3•3	Identify prime numbers by building all possible rectangular arrays for numbers.	Gain experience with the relationship between factors and prime numbers. [Number and Numeration Goal 3]
3•3	Play *Multiplication Top-It*.	Practice multiplication facts. [Operations and Computation Goal 3]
3•5	Complete Fact Triangles and write related fact families.	Gain experience with the connection between multiplication and division. [Operations and Computation Goal 3]
3•7	Solve *5-Minute Math* problems involving scale and measurements.	Practice using a map scale. [Operations and Computation Goal 7]
3•10	Take a 50-facts test.	Practice multiplication facts. [Operations and Computation Goal 3]
3•10	Solve *5-Minute Math* problems involving parentheses in number sentences.	Practice inserting parentheses in number sentences. [Patterns, Functions, and Algebra Goal 3]
3•11	Solve Broken-Calculator problems.	Practice solving open sentences. [Patterns, Functions, and Algebra Goal 2]

Looking at Grade-Level Goals

Everyday Mathematics develops concepts and skills over time. Below is a chart showing where the Grade-Level Goals emphasized in this unit are addressed throughout the year. Use the chart to help you determine which Maintaining Concepts and Skills activities on page 68 to utilize to ensure that students continue working toward these Grade-Level Goals.

Legend:
- ● Grade-Level Goal is taught.
- ◐ Grade-Level Goal is practiced and applied.
- ○ Grade-Level Goal is not a focus.

Grade-Level Goals Emphasized in Unit 3	Unit 1	2	3	4	5	6	7	8	9	10	11	12
[Number and Numeration Goal 5] Use numerical expressions to find and represent equivalent names for fractions and decimals; use and explain a multiplication rule to find equivalent fractions; rename fourths, fifths, tenths, and hundredths as decimals and percents.	○	○	●	●	○	○	◐	◐	●	◐	○	◐
[Operations and Computation Goal 1] Demonstrate automaticity with addition and subtraction fact extensions.	◐	●	◐	○	○	○	◐	◐	○	○	○	◐
[Operations and Computation Goal 2] Use manipulatives, mental arithmetic, paper-and-pencil algorithms and models, and calculators to solve problems involving the addition and subtraction of whole numbers and decimals through hundredths; describe the strategies used and explain how they work.	◐	◐	●	●	●	●	●	◐	◐	◐	◐	◐
[Operations and Computation Goal 3] Demonstrate automaticity with multiplication facts through 10 * 10 and proficiency with related division facts; use basic facts to compute fact extensions such as 30 * 60.	○	○	●	●	●	●	●	○	○	○	○	◐
[Operations and Computation Goal 4] Use manipulatives, mental arithmetic, paper-and-pencil algorithms and models, and calculators to solve problems involving the multiplication of multidigit whole numbers by 2-digit whole numbers and the division of multidigit whole numbers by 1-digit whole numbers; describe the strategies used and explain how they work.	○	○	●	○	●	●	◐	◐	○	○	○	◐
[Operations and Computation Goal 6] Make reasonable estimates for whole number and decimal addition and subtraction problems and whole number multiplication and division problems; explain how the estimates were obtained.	○	●	●	●	●	◐	◐	◐	●	◐	◐	◐
[Data and Chance Goal 2] Use the maximum, minimum, range, median, mode, and graphs to ask and answer questions, draw conclusions, and make predictions.	●	●	●	●	●	●	●	◐	◐	◐	◐	◐
[Patterns, Functions, and Algebra Goal 1] Extend, describe, and create numeric patterns; describe rules for patterns and use them to solve problems; use words and symbols to describe and write rules for functions that involve the four basic arithmetic operations and use those rules to solve problems.	◐	●	●	○	●	●	●	◐	○	◐	◐	◐
[Patterns, Functions, and Algebra Goal 2] Use conventional notation to write expressions and number sentences using the four basic arithmetic operations; determine whether number sentences are true or false; solve open sentences and explain the solutions; write expressions and number sentences to model number stories.	○	●	●	●	●	◐	◐	◐	◐	◐	◐	◐
[Patterns, Functions, and Algebra Goal 3] Evaluate numeric expressions containing grouping symbols; insert grouping symbols to make number sentences true.	○	○	●	◐	●	◐	◐	○	○	○	○	◐

Maintaining Concepts and Skills

Many of the goals addressed in this unit will be addressed again in later units. Those goals marked with an asterisk (*) are only addressed in future units as practice and application. Here are several suggestions for maintaining concepts and skills until they are formally revisited.

Operations and Computation Goal 1*

◆ Have students use their Addition/Subtraction Fact Triangles.

Operations and Computation Goal 3

◆ Have students play *Multiplication Top-It*.

◆ Have students make rectangular arrays with centimeter cubes. See the Readiness activity in Lesson 3-2 for more information.

◆ Have students use the number grid to explore skip-count patterns. See the Readiness activity in Lesson 3-3 for more information.

Patterns, Functions, and Algebra Goal 1

◆ Have students play *Seega*.

◆ Have students explore perimeter patterns. See the Readiness activity in Lesson 3-1 for more information.

◆ Use Frames-and-Arrows masters A and B on pages 142 and 143 of this handbook to create practice problems.

◆ Use the "What's My Rule?" master on page 144 to create practice problems.

Assessment

See page 70 in the *Assessment Handbook* for modifications to the written portion of the Unit 3 Progress Check.

Additionally, see pages 71–75 for modifications to the open-response task and selected student work samples.

Activities and Ideas for Differentiation

In this unit, students order, add, subtract, and make estimates for problems involving decimal computation. This section summarizes opportunities for supporting multiple learning styles and ability levels. Use these suggestions to develop a differentiation plan for Unit 4.

Part 1 Activities That Support Differentiation

Below are examples of Unit 4 activities that highlight some of the general instructional strategies that are hallmarks of a differentiated classroom. These strategies will help you support, emphasize, and enhance lesson content to make sure all of your students are engaged in the mathematics at the highest possible level. For more information about general differentiation strategies that accommodate the diverse needs of today's classrooms, see the essay on pages 8–16 of this handbook.

Lesson	Activity	Strategy
4◆1	Students skip count on their calculators to complete missing intervals on a number line.	Modeling physically
4◆2	Students model decimal numbers with base-10 blocks.	Modeling concretely
4◆3	Students use base-10 blocks to compare decimal numbers.	Modeling concretely
4◆4	Students explore decimals in the context of a trip odometer.	Making connections to everyday life
4◆6	Students perform decimal computation in the context of bank savings accounts.	Making connections to everyday life
4◆8	Students use a meterstick to explore the decimal relationships among metric linear units of measure.	Modeling visually

Vocabulary Development

The list below identifies the Key Vocabulary terms from this unit. The lesson in which each term is defined is indicated next to the term. Some of these terms or their homophones are used outside of mathematics. Consider adding other words as appropriate for developing understanding of the context of the lessons.

Lessons include suggestions for helping English language learners understand and develop vocabulary. For more information, see pages 17–19 of this handbook.

Key Vocabulary

*balance **4♦6**	†ONE **4♦2**
centimeter (cm) **4♦8**	personal measurement reference **4♦9**
decimal **4♦3**	speedometer **4♦4**
decimeter (dm) **4♦8**	tenth **4♦2**
*deposit **4♦6**	thousandth **4♦7**
hundredth **4♦2**	trip meter (*trip) **4♦4**
*interest **4♦6**	*unit **4♦2**
*meter (m) **4♦8**	*†whole **4♦2**
millimeter (mm) **4♦8**	withdrawal **4♦6**

★ Discuss the everyday and mathematical meanings of the words that are marked with an asterisk.

† For words marked with a dagger, write the words and their homophones on the board. For example, *ONE* and *won* and *whole* and *hole*. Discuss and clarify the meaning of each.

◆ As each word is introduced in the lesson, write the word on the board and discuss its meaning.

◆ List the words on a Math Word Wall for students to see. As each word is introduced in the lesson, add a picture next to the word on the Word Wall.

◆ Use the vocabulary words regularly when teaching lessons, and encourage students to use the words in their discussions.

 # Games

Below are suggested Unit 4 game adaptations. For more information about implementing games in a differentiated classroom, see pages 20–25 of this handbook.

Game: *Base-10 Exchange*

Skill Practiced: **Explore relationships in decimal place values.** [Number and Numeration Goal 1]

Modification	Purpose of Modification
Players use their base-10 grids on *Math Masters,* page 426. They fill their grid with cubes, trading 10 cubes for a long when they fill a column. They can also skip count by hundredths on the calculator to keep track of when their hundredths total a tenth.	Students explore relationships in decimal place values. [Number and Numeration Goal 1]
Players begin with a flat and subtract the number of cubes on each roll, exchanging as necessary.	Students subtract decimals. [Operations and Computation Goal 2]

Game: *Product Pile-Up*

Skill Practiced: **Solve multiplication facts.** [Operations and Computation Goal 3]

Modification	Purpose of Modification
Instead of two cards, players use one card. For a round, they select one factor by which everyone must multiply a card from his or her hand on each turn.	Students solve a group of multiplication facts with one factor the same. [Operations and Computation Goal 3]
Players can choose to make an extended fact by multiplying one factor by 10. If they get the incorrect answer for their extended fact, they draw two cards.	Students solve extended multiplication facts. [Operations and Computation Goal 3]

Game: *Number Top-It* (Decimals)

Skill Practiced: **Compare and order sets of decimals.** [Number and Numeration Goal 6]

Modification	Purpose of Modification
Players build their numbers with base-10 blocks on their base-10 grids (*Math Masters,* page 426) to compare the numbers.	Students compare decimals. [Number and Numeration Goal 6]
On each turn, a player makes two numbers. Their score is the difference between the two numbers. The highest score wins the round.	Students subtract and compare decimals. [Number and Numeration Goal 6; Operations and Computation Goal 2]

 # Math Boxes

Suggestions for using Math Boxes to meet individual needs begin on page 26 of this handbook. There are blank masters for Math Boxes on pages 134–139.

Using Part 3 of the Lessons

Use your professional judgment, along with assessment results, to determine whether the whole class, small groups, or individual students might benefit from these Unit 4 activities. Consider using the Part 3 Planning Master found on page 152 of this handbook to record your plans.

Readiness Activities

Lesson	Activity	Purpose of Activity
4•1	Make bill and coin combinations for given amounts.	Explore decimals to hundredths. [Number and Numeration Goal 1]
4•2	Play *Base-10 Exchange.*	Explore the relationship among hundredths, tenths, and ones. [Number and Numeration Goal 1]
4•3	Play *Coin Top-It.*	Gain experience comparing decimals in a money context. [Number and Numeration Goal 6]
4•4	Estimate the total cost of items.	Explore estimation with decimals. [Operations and Computation Goal 6]
4•5	Use a decimal version of the number grid to add and subtract decimals.	Explore place value, addition, and subtraction with decimals. [Number and Numeration Goal 1; Operations and Computation Goal 2]
4•6	Find totals and make change.	Explore adding and subtracting decimals. [Operations and Computation Goal 2]
4•7	Skip count by thousandths using calculators.	Explore extending the place-value system to thousandths. [Number and Numeration Goal 1]
4•8	Explore relationships among centimeters, decimeters, and meters.	Explore the relationships among metric units of linear measure. [Measurement and Reference Frames Goal 3]
4•9	Match metric units to measurements.	Explore relative sizes of metric units of linear measure. [Measurement and Reference Frames Goal 1]
4•10	Measure lengths of pencils with a centimeter ruler.	Explore the need for standard units of measure. [Measurement and Reference Frames Goal 1]

English Language Learners Support Activities

Lesson	Activity	Purpose of Activity
4•3	Create a *Decimals* All Around Museum.	Make connections between mathematics and everyday life; discuss new mathematical ideas. [Number and Numeration Goal 1]
4•6	Add *deposit, withdrawal,* and *balance* to the Math Word Bank.	Make connections between new terms and terms students know; use visual models to represent the terms. [Operations and Computation Goal 2]
4•8	Add *millimeter, centimeter,* and *meter* to the Math Word Bank.	Make connections between new terms and terms students know; use visual models to represent the terms. [Measurement and Reference Frames Goal 1]

Enrichment Activities

Lesson	Activity	Purpose of Activity
4•1	Create and solve place-value puzzles.	Apply understanding of place value. [Number and Numeration Goal 1]
4•2	Identify the values of base-10 blocks when different combinations are designated as the ONE.	Apply understanding of the whole or ONE. [Number and Numeration Goal 2]
4•3	Write and solve decimals riddles.	Apply understanding of decimal concepts. [Number and Numeration Goal 1]
4•4	Solve gasoline-mileage problems.	Apply understanding of estimating with decimals. [Operations and Computation Goal 6]
4•4	Solve a decimal magic-square puzzle.	Apply understanding of place value and addition of decimals. [Operations and Computation Goal 6]
4•5	Compute distances on a hiking trail.	Apply decimal-computation skills. [Operations and Computation Goal 2]
4•6	Compute the cost of the contents of "goodie bags."	Apply decimal-computation skills. [Operations and Computation Goal 2]
4•7	Analyze softball batting averages.	Apply understanding of decimals in the thousandths. [Number and Numeration Goal 6]
4•8	Explore the use of prefixes in metric units.	Explore metric units of linear measure. [Measurement and Reference Frames Goal 3]
4•9	Design a measurement scavenger hunt.	Apply understanding of metric units of linear measure. [Measurement and Reference Frames Goal 1]
4•10	Read and discuss the ratios in *If You Hopped Like a Frog*.	Explore the concept of scale. [Operations and Computation Goal 7]

Extra Practice Activities

Lesson	Activity	Purpose of Activity
4•1	Make and use a Compact Place-Value Flip Book.	Practice decimal place value. [Number and Numeration Goal 1]
4•3	Solve *5-Minute Math* problems involving decimals.	Gain experience with decimals. [Number and Numeration Goal 1]
4•7	Play *Base-10 Exchange*.	Practice exchanging tenths, hundredths, and thousandths. [Number and Numeration Goal 1]
4•9	Solve *5-Minute Math* problems involving metric measurements.	Gain experience with metric measurements. [Measurement and Reference Frames Goal 1]
4•10	Draw line segments and measure them to the nearest millimeter.	Practice measuring to the nearest millimeter. [Measurement and Reference Frames Goal 1]
4•10	Take a 50-facts test.	Practice multiplication facts. [Operations and Computation Goal 3]

Looking at Grade-Level Goals

Everyday Mathematics develops concepts and skills over time. Below is a chart showing where the Grade-Level Goals emphasized in this unit are addressed throughout the year. Use the chart to help you determine which Maintaining Concepts and Skills activities on page 75 to utilize to ensure that students continue working toward these Grade-Level Goals.

- ● Grade-Level Goal is taught.
- ◐ Grade-Level Goal is practiced and applied.
- ○ Grade-Level Goal is not a focus.

Grade-Level Goals Emphasized in Unit 4	1	2	3	4	5	6	7	8	9	10	11	12
[Number and Numeration Goal 1] Read and write whole numbers up to 1,000,000,000 and decimals through thousandths; identify places in such numbers and the values of the digits in those places; translate between whole numbers and decimals represented in words and in base-10 notation.	◐	●	●	●	●	◐	○	○	◐	○	○	○
[Number and Numeration Goal 2] Read, write, and model fractions; solve problems involving fractional parts of a region or a collection; describe and explain strategies used; given a fractional part of a region or a collection, identify the unit whole.	○	○	○	●	○	◐	●	○	●	◐	◐	◐
[Number and Numeration Goal 5] Use numerical expressions to find and represent equivalent names for fractions and decimals; use and explain a multiplication rule to find equivalent fractions; rename fourths, fifths, tenths, and hundredths as decimals and percents.	○	○	○	●	○	◐	●	◐	●	◐	◐	◐
[Number and Numeration Goal 6] Compare and order whole numbers up to 1,000,000,000 and decimals through thousandths; compare and order integers between −100 and 0; use area models, benchmark fractions, and analyses of numerators and denominators to compare and order fractions.	◐	◐	●	●	●	◐	●	◐	●	◐	◐	◐
[Operations and Computation Goal 2] Use manipulatives, mental arithmetic, paper-and-pencil algorithms and models, and calculators to solve problems involving the addition and subtraction of whole numbers and decimals through hundredths; describe the strategies used and explain how they work.	◐	●	●	●	●	◐	◐	◐	◐	◐	◐	◐
[Operations and Computation Goal 3] Demonstrate automaticity with multiplication facts through 10 * 10 and proficiency with related division facts; use basic facts to compute fact extensions such as 30 * 60.	○	●	●	●	●	●	○	○	○	◐	◐	◐
[Operations and Computation Goal 6] Make reasonable estimates for whole number and decimal addition and subtraction problems and whole number multiplication and division problems; explain how the estimates were obtained.	○	○	●	●	●	◐	◐	◐	◐	◐	◐	◐
[Measurement and Reference Frames Goal 1] Estimate length with and without tools; measure length to the nearest $\frac{1}{4}$ inch and $\frac{1}{2}$ centimeter; use tools to measure and draw angles; estimate the size of angles without tools.	●	●	◐	◐	○	◐	◐	◐	◐	◐	◐	●
[Measurement and Reference Frames Goal 3] Describe relationships among U.S. customary units of measure and among metric units of measure.	◐	◐	◐	◐	◐	◐	◐	●	◐	◐	◐	◐

Unit

Maintaining Concepts and Skills

All of the goals addressed in this unit will be addressed again in later units. Here are several suggestions for maintaining concepts and skills until they are formally revisited.

Number and Numeration Goal 1

◆ Have students play *Base-10 Exchange* and *Fishing for Digits*.

◆ Have students build numbers with base-10 blocks.

Number and Numeration Goal 6

◆ Have students play *Coin Top-It* and *Number Top-It*.

Operations and Computation Goal 2

◆ Have students find the total cost for items and compute change. See the Readiness activity in Lesson 4-6 for more information.

◆ Have students use a decimal version of the number grid to add and subtract decimals. See the Readiness activity in Lesson 4-5 for more information.

Operations and Computation Goal 3

◆ Have students play *Product Pile-Up*.

◆ Have students use their Multiplication/Division Fact Triangles.

◆ Use the Number Grid master on page 146 in this handbook and have students shade multiples to see patterns.

◆ Use Frames-and-Arrows masters A and B on pages 142 and 143 of this handbook to have students practice with skip counting using addition rules.

Assessment

See page 78 in the *Assessment Handbook* for modifications to the written portion of the Unit 4 Progress Check.

Additionally, see pages 79–83 for modifications to the open-response task and selected student work samples.

Activities and Ideas for Differentiation

In this unit, students extend their skills with pencil-and-paper multiplication algorithms and explore ordering and comparing large numbers. This section summarizes opportunities for supporting multiple learning styles and ability levels. Use these suggestions to develop a differentiation plan for Unit 5.

Part 1 Activities That Support Differentiation

Below are examples of Unit 5 activities that highlight some of the general instructional strategies that are hallmarks of a differentiated classroom. These strategies will help you support, emphasize, and enhance lesson content to make sure all of your students are engaged in the mathematics at the highest possible level. For more information about general differentiation strategies that accommodate the diverse needs of today's classrooms, see the essay on pages 8–16 of this handbook.

Lesson	Activity	Strategy
5◆1	Students use base-10-block arrays to demonstrate one solution strategy for an extended-multiplication-facts problem.	Modeling concretely
5◆2	Students use a record sheet to organize a partial-products multiplication algorithm.	Using organizational tools
5◆3	Students estimate sums in the context of travel distances on a map.	Making connections to everyday life
5◆6, 5◆7	Students use partial-products and lattice algorithms to solve multiplication problems.	Incorporating and validating a variety of methods
5◆8	Students estimate the total number of dots arranged in a large array which is subdivided into smaller arrays.	Modeling visually
5◆10	Students use a table to record the steps of a rounding strategy.	Using organizational tools

Vocabulary Development

The list below identifies the Key Vocabulary terms from this unit. The lesson in which each term is defined is indicated next to the term. Some of these terms or their homophones are used outside of mathematics. Consider adding other words as appropriate for developing understanding of the context of the lessons.

Lessons include suggestions for helping English language learners understand and develop vocabulary. For more information, see pages 17–19 of this handbook.

Key Vocabulary

billion **5◆8**

estimation **5◆3**

exponent **5◆9**

extended multiplication facts **5◆1**

*lattice **5◆7**

lattice method (for multiplication) **5◆7**

magnitude estimate **5◆4**

million **5◆8**

partial product (*product) **5◆5**

partial-products method **5◆5**

powers of 10 (*power) **5◆9**

quadrillion **5◆9**

quintillion **5◆9**

rough estimate (†rough) **5◆4**

*round **5◆3**

rounding (to a certain place) **5◆10**

scientific notation **5◆9**

sextillion **5◆9**

trillion **5◆9**

* Discuss the everyday and mathematical meanings of the words that are marked with an asterisk.

† For the word marked with a dagger, write *rough* and its homophone *ruff* on the board. Discuss and clarify the meaning of each.

◆ As each word is introduced in the lesson, write the word on the board and discuss its meaning.

◆ List the words on a Math Word Wall for students to see. As each word is introduced in the lesson, add a picture next to the word on the Word Wall.

◆ Use the vocabulary words regularly when teaching lessons, and encourage students to use the words in their discussions.

 **Games**

Below are suggested Unit 5 game adaptations. For more information about implementing games in a differentiated classroom, see pages 20–25 of this handbook.

Game: *Multiplication Wrestling*

Skill Practiced: **Calculate and find the sum of partial products.** [Operations and Computation Goal 4; Patterns, Functions, and Algebra Goal 4]

Modification	Purpose of Modification
Players draw three cards and make one 2-digit number. They use the remaining card to determine what multiple of 10 to multiply by.	Students calculate and find the sum of partial products. [Operations and Computation Goal 4; Patterns, Functions, and Algebra Goal 4]
Players can choose to form two 2-digit numbers or one 3-digit number and one 1-digit number for their factors.	Students calculate and find the sum of partial products. [Operations and Computation Goal 4; Patterns, Functions, and Algebra Goal 4]

Game: *Name That Number*

Skill Practiced: **Find equivalent names for numbers.** [Number and Numeration Goal 4]

Modification	Purpose of Modification
Players keep the same target number for each round; for example, 10 is always the target number.	Students find equivalent names for one target number. [Number and Numeration Goal 4]
The card turned over for the target is multiplied by 10 to find the target number; for example, an 8 is turned over so the target is 80.	Students find equivalent names for multiples of 10. [Number and Numeration Goal 4]

Game: *High-Number Toss*

Skill Practiced: **Identify value of digits and compare large numbers.** [Number and Numeration Goals 1 and 6]

Modification	Purpose of Modification
Players use a place-value chart (*Math Masters,* page 398) to record their numbers.	Students identify values of digits and compare large numbers within the structure of a place-value chart. [Number and Numeration Goal 1]
Players receive a bonus point if they can write their number as a 3-digit number times some power of 10 written in exponential notation.	Students identify the value of digits. [Number and Numeration Goal 1]

 **Math Boxes**

Suggestions for using Math Boxes to meet individual needs begin on page 26 of this handbook. There are blank masters for Math Boxes on pages 134–139.

Using Part 3 of the Lessons

Use your professional judgment, along with assessment results, to determine whether the whole class, small groups, or individual students might benefit from these Unit 5 activities. Consider using the Part 3 Planning Master found on page 152 of this handbook to record your plans.

Readiness Activities

Lesson	Activity	Purpose of Activity
5◆1	Play *Multiplication Top-It.*	Gain experience with basic multiplication facts. [Operations and Computation Goal 3]
5◆2	Solve addition problems using partial-sums addition.	Explore adding multidigit numbers in preparation for using a partial-products multiplication algorithm. [Operations and Computation Goal 2]
5◆3	Round numbers building them first with base-10 blocks.	Gain experience with estimation skills. [Operations and Computation Goal 6]
5◆4	Plot whole numbers on a curved number line.	Explore rounding whole numbers. [Operations and Computation Goal 6]
5◆5	Model multiplication problems with base-10 blocks.	Explore 1-digit times 2-digit multiplication. [Operations and Computation Goal 4]
5◆5	Record related sets of extended facts and use base-10 blocks to identify patterns.	Explore patterns in extended facts. [Operations and Computation Goal 3]
5◆6	Model multiplication problems with base-10 blocks.	Explore 2-digit times 2-digit multiplication. [Operations and Computation Goal 4]
5◆7	Find patterns in the Fact Lattice.	Explore the use of the lattice grid for multiplication. [Operations and Computation Goal 4]
5◆8	Play *High-Number Toss.*	Gain experience with place-value skills. [Number and Numeration Goals 1 and 6]
5◆10	Identify numbers halfway between large numbers on a number line.	Gain experience locating large numbers on a number line. [Operations and Computation Goal 6]
5◆11	Play *Number Top-It* on a place-value mat.	Explore comparing large numbers. [Number and Numeration Goal 6]

English Language Learners Support Activities

Lesson	Activity	Purpose of Activity
5◆1	Add *extended fact* to the Math Word Bank.	Make connections between a new term and terms students know; use visual models to represent the term. [Operations and Computation Goal 3]
5◆3	Add *estimation* and *round* to the Math Word Bank.	Make connections between new terms and terms students know; use visual models to represent the terms. [Operations and Computation Goal 6]
5◆7	Make posters to illustrate the various *multiplication algorithms.*	Use a student-made poster as a visual reference for the algorithms. [Operations and Computation Goal 4]
5◆9	Brainstorm uses of the word *power.*	Clarify the mathematical and everyday uses of the term. [Number and Numeration Goal 4]

Enrichment Activities

Lesson	Activity	Purpose of Activity
5·2	Judge a *Multiplication Wrestling* competition using estimation strategies.	Apply understanding of the Distributive Property of Multiplication over Addition. [Patterns, Functions, and Algebra Goal 4]
5·3	Use estimation skills to find the shortest route between four cities.	Apply understanding of estimating sums. [Operations and Computation Goal 6]
5·4	Find missing numbers and digits in multiplication number sentences.	Apply understanding of estimates. [Operations and Computation Goal 6]
5·5	Solve a multiplication puzzle.	Apply multiplication skills. [Operations and Computation Goal 3]
5·6	Solve a multistep number story about scoring a dart game.	Apply multidigit-multiplication skills. [Operations and Computation Goal 4]
5·6	Complete Venn diagrams based on factors.	Apply understanding of extended multiplication and division facts. [Operations and Computation Goal 3]
5·6	Write multiplication number stories.	Apply understanding of multiplication algorithms. [Operations and Computation Goal 4]
5·7	Solve problems using Napier's Rods.	Apply understanding of lattice multiplication. [Operations and Computation Goal 4]
5·8	Estimate the number of dots on a paper and the weight of the paper.	Apply understanding of the relationships among thousands, millions, and billions. [Number and Numeration Goal 1]
5·8	Read *How Much Is a Million?* and discuss the large numbers in the story.	Explore understanding of the relationships among thousands, millions, billions, and trillions. [Number and Numeration Goal 1]
5·9	Identify and discuss patterns in calculator displays for powers of 10.	Explore exponential notation. [Number and Numeration Goal 4]
5·10	Round bar-graph data involving populations.	Apply understanding of rounding data. [Operations and Computation Goal 6]

Extra Practice Activities

Lesson	Activity	Purpose of Activity
5·1	Solve multiplication/division puzzles.	Practice with extended facts. [Operations and Computation Goal 3]
5·3	Solve elapsed-time problems.	Practice using a measurement scale to determine elapsed time. [Measurement and Reference Frames Goal 3]
5·4	Solve *5-Minute Math* problems involving estimating products.	Practice using estimation skills. [Operations and Computation Goal 6]
5·9	Solve *5-Minute Math* problems involving exponential notation.	Practice with exponential notation. [Number and Numeration Goal 4]
5·10	Take a 50-facts test.	Practice multiplication facts. [Operations and Computation Goal 3]
5·10	Solve *5-Minute Math* problems involving rounding.	Practice rounding. [Operations and Computation Goal 6]
5·11	Play *High-Number Toss.*	Practice comparing large numbers. [Number and Numeration Goals 1 and 6]

Looking at Grade-Level Goals

Everyday Mathematics develops concepts and skills over time. Below is a chart showing where the Grade-Level Goals emphasized in this unit are addressed throughout the year. Use the chart to help you determine which Maintaining Concepts and Skills activities on page 82 to utilize to ensure that students continue working toward these Grade-Level Goals.

- ● Grade-Level Goal is taught.
- ◐ Grade-Level Goal is practiced and applied.
- ○ Grade-Level Goal is not a focus.

Grade-Level Goals Emphasized in Unit 5	1	2	3	4	5	6	7	8	9	10	11	12
[Number and Numeration Goal 1] Read and write whole numbers up to 1,000,000,000 and decimals through thousandths; identify places in such numbers and the values of the digits in those places; translate between whole numbers and decimals represented in words and in base-10 notation.	◐	●	◐	◐	◐	◐	○	○	◐	○	○	○
[Number and Numeration Goal 4] Use numerical expressions involving one or more of the basic four arithmetic operations and grouping symbols to give equivalent names for whole numbers.	◐	●	●	●	◐	◐	○	◐	◐	○	○	○
[Operations and Computation Goal 2] Use manipulatives, mental arithmetic, paper-and-pencil algorithms and models, and calculators to solve problems involving the addition and subtraction of whole numbers and decimals through hundredths; describe the strategies used and explain how they work.	◐	●	●	●	●	◐	◐	◐	◐	◐	◐	◐
[Operations and Computation Goal 3] Demonstrate automaticity with multiplication facts through 10 * 10 and proficiency with related division facts; use basic facts to compute fact extensions such as 30 * 60.	○	○	●	●	●	●	○	○	○	○	○	○
[Operations and Computation Goal 4] Use manipulatives, mental arithmetic, paper-and-pencil algorithms and models, and calculators to solve problems involving the multiplication of multidigit whole numbers by 2-digit whole numbers and the division of multidigit whole numbers by 1-digit whole numbers; describe the strategies used and explain how they work.	○	○	○	○	●	○	○	●	●	○	○	○
[Operations and Computation Goal 6] Make reasonable estimates for whole number and decimal addition and subtraction problems and whole number multiplication and division problems; explain how the estimates were obtained.	○	●	●	●	●	●	●	●	◐	◐	○	○
[Data and Chance Goal 2] Use the maximum, minimum, range, median, mode, and graphs to ask and answer questions, draw conclusions, and make predictions.	◐	●	●	●	●	○	○	○	○	○	○	○
[Patterns, Functions, and Algebra Goal 1] Extend, describe, and create numeric patterns; describe rules for patterns and use them to solve problems; use words and symbols to describe and write rules for functions that involve the four basic arithmetic operations and use those rules to solve problems.	◐	●	●	◐	●	○	○	○	○	○	○	○
[Patterns, Functions, and Algebra Goal 4] Describe and apply the Distributive Property of Multiplication over Addition.	○	○	○	○	○	○	○	○	○	○	○	○

Maintaining Concepts and Skills

Some of the goals addressed in this unit will be addressed again in later units. Those goals marked with an asterisk (*) are only addressed in future units as practice and application. Here are several suggestions for maintaining concepts and skills until they are formally revisited.

Number and Numeration Goal 1*

◆ Have students play *High-Number Toss*.

◆ Use the "What's My Rule?" master on page 144 of this handbook to create place-value practice problems with rules such as $+10$, -10, $+100$, -100, and so on.

Number and Numeration Goal 4*

◆ Have students play *Name That Number*.

Operations and Computation Goal 4

◆ Have students model multiplication with base-10 blocks. See the Readiness activities in Lessons 5-5 and 5-6.

◆ Have students find patterns in the Fact Lattice. See the Readiness activity in Lesson 5-7 for more information.

Operations and Computation Goal 6

◆ Have students explore rounding using base-10 blocks and on a curved number line. See the Readiness activities in Lessons 5-3 and 5-4.

Patterns, Functions, and Algebra Goal 4*

◆ Have students play *Multiplication Wrestling*.

Assessment

See page 86 in the *Assessment Handbook* for modifications to the written portion of the Unit 5 Progress Check.

Additionally, see pages 87–90 for modifications to the open-response task and selected student work samples.

Activities and Ideas for Differentiation

In this unit, students explore a partial-quotients division algorithm, coordinate-grid systems, and measuring angles. This section summarizes opportunities for supporting multiple learning styles and ability levels. Use these suggestions to develop a differentiation plan for Unit 6.

Part 1 Activities That Support Differentiation

Below are examples of Unit 6 activities that highlight some of the general instructional strategies that are hallmarks of a differentiated classroom. These strategies will help you support, emphasize, and enhance lesson content to make sure all of your students are engaged in the mathematics at the highest possible level. For more information about general differentiation strategies that accommodate the diverse needs of today's classrooms, see the essay on pages 8–16 of this handbook.

Lesson	Activity	Strategy
6•2	Use a multiplication/division diagram to organize the information in a number story.	Using organizational tools
6•4	Students draw pictures to model division problems.	Modeling visually
6•5	Use a clock face to model angles of different sizes and compare the angle sizes.	Modeling visually
6•5	Students use a straw to demonstrate the meaning and sizes of angles.	Modeling concretely
6•8	Students estimate distances on a map that are not linear by counting squares, using a compass to measure, and using the edge of a sheet of paper.	Modeling physically
6•10	Model division using base-10 blocks.	Modeling concretely

Vocabulary Development

The list below identifies the Key Vocabulary terms from this unit. The lesson in which each term is defined is indicated next to the term. Some of these terms or their homophones are used outside of mathematics. Consider adding other words as appropriate for developing understanding of the context of the lessons.

Lessons include suggestions for helping English language learners understand and develop vocabulary. For more information, see pages 17–19 of this handbook.

Key Vocabulary

acute angle **6♦7**	hemisphere **6♦9**	prime meridian **6♦9**
*angle (∠) **6♦6**	index of locations **6♦8**	quotient **6♦2**
axis **6♦9**	latitude (lines) **6♦9**	reflex angle (*reflex) **6♦7**
base line (*†base) **6♦7**	letter-number pair (†pair) **6♦8**	remainder **6♦2**
clockwise **6♦5**	longitude (lines) **6♦9**	right angle (*†right) **6♦5**
clockwise rotation **6♦6**	map scale (*scale) **6♦8**	rotation **6♦5**
counterclockwise rotation **6♦6**	meridian bar (*bar) **6♦9**	sides (of an angle) (†side) **6♦6**
*degree **6♦5**	mixed number (*mixed) **6♦4**	South Pole **6♦9**
dividend **6♦3**	Multiplication/Division Diagram **6♦1**	sphere **6♦9**
†divisor **6♦3**	North Pole **6♦9**	straight angle (†straight) **6♦7**
equal-groups notation **6♦2**	obtuse angle **6♦7**	*†turn **6♦5**
equator **6♦9**	ordered number pair (†pair) **6♦8**	vertex (of an angle) **6♦6**
full-circle protractor **6♦6**	parallels **6♦9**	
half-circle protractor **6♦7**	partial quotient **6♦3**	

* Discuss the everyday and mathematical meanings of the words that are marked with an asterisk.

† For words marked with a dagger, write the words and their homophones on the board. For example, *base* and *bass; divisor* and *deviser; pair, pare,* and *pear; right, rite, wright,* and *write; side* and *sighed; straight* and *strait;* and *turn* and *tern.* Discuss and clarify the meaning of each.

♦ As each word is introduced in the lesson, write the word on the board and discuss its meaning.

♦ List the words on a Math Word Wall for students to see. As each word is introduced in the lesson, add a picture next to the word on the Word Wall.

♦ Use the vocabulary words regularly when teaching lessons, and encourage students to use the words in their discussions.

 **Games**

Below are suggested Unit 6 game adaptations. For more information about implementing games in a differentiated classroom, see pages 20–25 of this handbook.

Game: *Beat the Calculator*

Skill Practiced: Multiply extended facts. [Operations and Computation Goal 3]

Modification	Purpose of Modification
Players keep one factor constant throughout the game. For example, the caller draws one card for each round and always multiplies that number by 30.	Students multiply a set of extended facts with one factor in common. [Operations and Computation Goal 3]
Players need three index cards labeled 10, 100, and 1,000. The caller first randomly selects an index card and then draws two number cards. The caller can choose to make one or both of the factors a multiple of the number on the index card.	Students multiply extended facts. [Operations and Computation Goal 3]

Game: *Division Dash*

Skill Practiced: Divide 2- or 3-digit dividends by 1-digit divisors. [Operations and Computation Goal 4]

Modification	Purpose of Modification
Players include only the 2 and 5 cards in the divisor pile.	Students divide 2-digit dividends using only 2 and 5 as divisors. [Operations and Computation Goal 4]
Players make a 3- or 4-digit dividend and the divisor a multiple of 10.	Students divide 3- or 4-digit dividends by divisors that are multiples of 10. [Operations and Computation Goal 4]

Game: *Angle Tangle*

Skill Practiced: Estimate and measure angles. [Measurement and Reference Frames Goal 1]

Modification	Purpose of Modification
Players estimate whether angles are larger or smaller than a right angle (90°). They can use the corner of a piece of paper to check their estimates.	Students compare angles to 90° angles and measure angles. [Measurement and Reference Frames Goal 1]
Players draw and measure reflex angles.	Students draw, estimate, and measure angles greater than 180°. [Measurement and Reference Frames Goal 1]

 **Math Boxes**

Suggestions for using Math Boxes to meet individual needs begin on page 26 of this handbook. There are blank masters for Math Boxes on pages 134–139.

Using Part 3 of the Lessons

Use your professional judgment, along with assessment results, to determine whether the whole class, small groups, or individual students might benefit from these Unit 6 activities. Consider using the Part 3 Planning Master found on page 152 of this handbook to record your plans.

Readiness Activities

Lesson	Activity	Purpose of Activity
6◆1	Play *Division Arrays*.	Explore the relationship between multiplication and division. [Operations and Computation Goal 7]
6◆2	Complete Fact Triangles with extended facts.	Explore the relationship between extended multiplication and division facts. [Operations and Computation Goal 3]
6◆3	Play *Beat the Calculator* with extended facts.	Gain experience multiplying extended facts. [Operations and Computation Goal 3]
6◆4	Read a *Remainder of One*, create arrays with cubes, and record division number models.	Explore the concept of remainders. [Operations and Computation Goal 3]
6◆5	Match times expressed in analog, digital, and word form.	Explore alternate ways of naming time. [Measurement and Reference Frames Goal 1]
6◆6	Make and use a waxed-paper protractor.	Explore the use of a protractor to measure angles. [Measurement and Reference Frames Goal 1]
6◆7	Model specified types of angles with rope.	Explore estimating angle measures. [Measurement and Reference Frames Goal 1]
6◆8	Describe the location of objects and place objects on a life-size coordinate grid.	Gain experience locating ordered pairs on a coordinate grid. [Measurement and Reference Frames Goal 4]
6◆10	Play *Division Dash*.	Gain experience with 2- or 3-digit dividends and 1-digit divisors. [Operations and Computation Goal 4]

English Language Learners Support Activities

Lesson	Activity	Purpose of Activity
6◆3	Display division number models and label the *dividend, divisor, quotient,* and *remainder* for each.	Use a student-made poster as a visual reference for new terms. [Operations and Computation Goal 4]
6◆5	Add *degree* to the Math Word Bank.	Make connections between a new term and terms students know; use visual models to represent the term. [Measurement and Reference Frames Goal 1]
6◆6	Discuss *clockwise* and *counterclockwise* as directions of angle rotation.	Clarify the mathematical and everyday uses of the terms. [Geometry Goal 3]
6◆7	Make a graphic organizer for *angle* using words and pictures.	Connect a new term to existing vocabulary; use a graphic organizer to describe the characteristics of angles. [Geometry Goal 1]

Enrichment Activities

Lesson	Activity	Purpose of Activity
6◆1	Write multiplication and division number stories.	Apply understanding of the inverse relationship between multiplication and division. [Operations and Computation Goal 3]
6◆3	Use a series of related clues to solve a number story involving division.	Apply understanding of division. [Operations and Computation Goal 4]
6◆4	Solve a division number story by finding multiples.	Apply understanding of multiples, factors, and division with remainders. [Number and Numeration Goal 3]
6◆5	Determine elapsed time for 1° increments on a clock face.	Explore the relationship between elapsed time and angle measures. [Geometry Goal 3]
6◆6	Play *Angle Add-Up*.	Apply understanding of addition and subtraction to find the measures of unknown angles. [Operations and Computation Goal 2]
6◆7	Measure angles and determine that the sum of the angle measures of any triangle is 180°.	Apply understanding of measuring angles. [Measurement and Reference Frames Goal 1]
6◆7	Read *Sir Cumference and the Great Knight of Angleland* and record what Radius learned on his quest.	Explore using the half-circle protractor and measuring angles. [Measurement and Reference Frames Goal 1]
6◆8	Play *Grid Search*.	Apply understanding of coordinate grids. [Measurement and Reference Frames Goal 4]
6◆9	Read *Sea Clocks: The Story of Longitude* and explain how sailors determine longitude at sea.	Apply understanding of longitude. [Measurement and Reference Frames Goal 4]
6◆10	Perform, analyze, and explain a division "magic trick."	Explore the concept of division using a calculator. [Operations and Computation Goal 4]

Extra Practice Activities

Lesson	Activity	Purpose of Activity
6◆2	Play *Buzz* and *Bizz-Buzz*.	Practice naming multiples of numbers. [Number and Numeration Goal 3]
6◆3	Play *Division Dash*.	Practice dividing 2- or 3-digit dividends by 1-digit divisors. [Operations and Computation Goal 4]
6◆4	Solve *5-Minute Math* problems involving division.	Practice solving division problems. [Operations and Computation Goal 4]
6◆5	Play *Robot*.	Practice making rotations expressed as fractions of turns and as degree measures. [Geometry Goal 3]
6◆6	Play *Angle Tangle*.	Practice estimating and measuring angles. [Measurement and Reference Frames Goal 1]
6◆8	Plot and name points on a coordinate grid.	Practice coordinate-grid skills. [Measurement and Reference Frames Goal 4]
6◆9	Use longitude and latitude to locate places on a map.	Practice using latitude and longitude. [Measurement and Reference Frames Goal 4]
6◆10	Take a 50-facts test.	Practice multiplication facts. [Operations and Computation Goal 3]

Looking at Grade-Level Goals

Everyday Mathematics develops concepts and skills over time. Below is a chart showing where the Grade-Level Goals emphasized in this unit are addressed throughout the year. Use the chart to help you determine which Maintaining Concepts and Skills activities on page 89 to utilize to ensure that students continue working toward these Grade-Level Goals.

Legend:
- ● Grade-Level Goal is taught.
- ◐ Grade-Level Goal is practiced and applied.
- ○ Grade-Level Goal is not a focus.

Grade-Level Goals Emphasized in Unit 6	1	2	3	4	5	6	7	8	9	10	11	12
[Number and Numeration Goal 3] Find multiples of whole numbers less than 10; identify prime and composite numbers; find whole-number factors of numbers.	○	○	○	○	○	●	◐	◐	◐	◐	◐	◐
[Operations and Computation Goal 1] Demonstrate automaticity with addition and subtraction fact extensions.	◐	●	◐	○	○	◐	◐	◐	○	○	○	◐
[Operations and Computation Goal 2] Use manipulatives, mental arithmetic, paper-and-pencil algorithms and models, and calculators to solve problems involving the addition and subtraction of whole numbers and decimals through hundredths; describe the strategies used and explain how they work.	◐	◐	●	●	●	◐	●	●	◐	◐	●	●
[Operations and Computation Goal 3] Demonstrate automaticity with multiplication facts through 10 * 10 and proficiency with related division facts; use basic facts to compute fact extensions such as 30 * 60.	○	○	◐	●	●	◐	●	●	○	◐	◐	◐
[Operations and Computation Goal 4] Use manipulatives, mental arithmetic, paper-and-pencil algorithms and models, and calculators to solve problems involving the multiplication of multidigit whole numbers by 2-digit whole numbers and the division of multidigit whole numbers by 1-digit whole numbers; describe the strategies used and explain how they work.	○	○	◐	●	●	◐	●	◐	○	○	○	◐
[Operations and Computation Goal 7] Use repeated addition, skip counting, arrays, area, and scaling to model multiplication and division.	○	○	◐	◐	◐	◐	◐	◐	○	◐	○	◐
[Measurement and Reference Frames Goal 1] Estimate length with and without tools; measure length to the nearest $\frac{1}{4}$ inch and $\frac{1}{2}$ centimeter; use tools to measure and draw angles; estimate the size of angles without tools.	●	●	◐	◐	◐	◐	●	●	◐	◐	◐	◐
[Measurement and Reference Frames Goal 4] Use ordered pairs of numbers to name, locate, and plot points in the first quadrant of a coordinate grid.	◐	●	◐	◐	◐	●	◐	◐	○	◐	○	◐
[Geometry Goal 1] Identify, draw, and describe points, intersecting and parallel line segments and lines, rays, and right, acute, and obtuse angles.	○	○	○	○	○	●	◐	◐	○	◐	○	◐
[Geometry Goal 2] Describe, compare, and classify plane and solid figures, including polygons, circles, spheres, cylinders, rectangular prisms, cones, cubes, and pyramids, using appropriate geometric terms including *vertex, base, face, edge,* and *congruent.*	○	○	○	○	○	●	◐	◐	◐	●	●	◐
[Geometry Goal 3] Identify, describe, and sketch examples of reflections; identify and describe examples of translations and rotations.	○	○	○	○	○	◐	○	○	◐	●	●	◐
[Patterns, Functions, and Algebra Goal 2] Use conventional notation to write expressions and number sentences using the four basic arithmetic operations; determine whether number sentences are true or false; solve open sentences and explain the solutions; write expressions and number sentences to model number stories.	○	○	◐	◐	◐	●	◐	◐	◐	◐	◐	●
[Patterns, Functions, and Algebra Goal 3] Evaluate numeric expressions containing grouping symbols; insert grouping symbols to make number sentences true.	○	○	◐	◐	◐	●	○	○	◐	◐	◐	●

Maintaining Concepts and Skills

Some of the goals addressed in this unit will be addressed again in later units. Those goals marked with an asterisk (*) are addressed in future units only as practice and application. Here are several suggestions for maintaining concepts and skills until goals are revisited.

Operations and Computation Goal 1*

◆ Have students play *Addition Top-It* and *Subtraction Top-It*.

◆ Have students practice with their Addition/Subtraction Fact Triangles.

◆ Use Frames-and-Arrows masters A and B on pages 142 and 143, the "What's My Rule?" master on page 144, and the Math Boxes C master on page 136. You can make problem sets for each of these routines that practice addition and subtraction facts.

Operations and Computation Goal 7

◆ Have students play *Division Arrays*.

◆ Have students use multiplication/division diagrams to solve number stories.

Measurement and Reference Frames Goal 1

◆ Have students play *Angle Tangle*.

◆ Have students make and use a waxed-paper protractor. See the Readiness activity in Lesson 6-6 for more information.

◆ Have students model specified types of angles with rope. See the Readiness activity in Lesson 6-7 for more information.

Measurement and Reference Frames Goal 4*

◆ Have students play *Grid Search*.

◆ Have students describe the location of objects and place objects on a life-size coordinate grid. See the Readiness activity in Lesson 6-8 for more information.

Patterns, Functions, and Algebra Goal 2*

◆ Have students play *Name That Number* and record number sentences, with parentheses where appropriate, to record their solutions.

Assessment

See page 94 in the *Assessment Handbook* for modifications to the written portion of the Unit 6 Progress Check.

Additionally, see pages 95–99 for modifications to the open-response task and selected student work samples.

Activities and Ideas for Differentiation

In this unit, students review fraction ideas introduced in earlier grades, further develop an understanding of equivalent fractions, and explore probability through informal activities. This section summarizes opportunities for supporting multiple learning styles and ability levels. Use these suggestions to develop a differentiation plan for Unit 7.

Part 1 Activities That Support Differentiation

Below are examples of Unit 7 activities that highlight some of the general instructional strategies that are hallmarks of a differentiated classroom. These strategies will help you support, emphasize, and enhance lesson content to make sure all of your students are engaged in the mathematics at the highest possible level. For more information about general differentiation strategies that accommodate the diverse needs of today's classrooms, see the essay on pages 8–16 of this handbook.

Lesson	Activity	Strategy
7♦1	Students use an area model with pattern blocks and a number-line model to identify fractional parts.	Modeling concretely; modeling visually
7♦2	Students use pennies to model and solve "fraction-of-a-collection" problems.	Modeling concretely
7♦5	Students use pattern blocks to model equivalent fractions and fraction and mixed-number addition and subtraction.	Modeling concretely
7♦8	Students use a 100-grid representation to explore relationships between fraction and decimal names for parts of a whole.	Modeling visually
7♦9	Students use area models on Fraction Cards to order and compare fractions.	Modeling visually
7♦11	Students use spinners to explore the relationship between the fraction of an area shaded and the probability of landing on that area.	Building on prior knowledge
7♦12a	Students use a visual fraction model to multiply fractions by whole numbers.	Modeling visually

Vocabulary Development

The list below identifies the Key Vocabulary terms from this unit. The lesson in which each term is defined is indicated next to the term. Some of these terms or their homophones are used outside of mathematics. Consider adding other words as appropriate for developing understanding of the context of the lessons.

Lessons include suggestions for helping English language learners understand and develop vocabulary. For more information, see pages 17–19 of this handbook.

Key Vocabulary

denominator 7♦1

equal chance 7♦11

equally likely 7♦3

equally (more, less) likely (†more) 7♦11

equation 7♦12a

equivalent fractions 7♦7

Equivalent Fractions Rule 7♦7

*event 7♦3

expect 7♦11

fair die (*†fair, *†die) 7♦11

favorable outcome 7♦3

mixed number (*mixed) 7♦1

multiple 7♦12a

numerator 7♦1

outcome 7♦3

probability 7♦3

†whole (or ONE or unit) (†ONE) 7♦1

"whole" box (*box) 7♦1

---

* Discuss the everyday and mathematical meanings of the words that are marked with an asterisk.

† For words marked with a dagger, write the words and their homophones on the board. For example, *more* and *moor; fair* and *fare; die* and *dye; whole* and *hole;* and *ONE* and *won.* Discuss and clarify the meaning of each.

♦ As each word is introduced in the lesson, write the word on the board and discuss its meaning.

♦ List the words on a Math Word Wall for students to see. As each word is introduced in the lesson, add a picture next to the word on the Word Wall.

♦ Use the vocabulary words regularly when teaching lessons, and encourage students to use the words in their discussions.

 **Games**

Below are suggested Unit 7 game adaptations. For more information about implementing games in a differentiated classroom, see pages 20–25 of this handbook.

Game: *Over and Up Squares*

Skill Practiced: Plot ordered number pairs on a coordinate grid. [Measurement and Reference Frames Goal 4]

Modification	Purpose of Modification
Players do not connect the points with line segments. At the end of the game, players get a bonus point for every segment they can draw between two adjacent points in their color.	Students plot ordered number pairs on a coordinate grid. [Measurement and Reference Frames Goal 4]
When doubles are rolled, a player can call out any ordered pair before plotting the point. If plotted incorrectly, the other player gets the point.	Students identify and plot ordered number pairs on a coordinate grid. [Measurement and Reference Frames Goal 4]

Game: *Fraction Match*

Skill Practiced: Identify equivalent fractions. [Number and Numeration Goal 5]

Modification	Purpose of Modification
Players use only the halves, fourths, eighths, twelfths, and WILD cards.	Students identify easier equivalent fractions. [Number and Numeration Goal 5]
Players match only equivalent fractions. When no more cards can be played, players show their cards and take turns finding a "match" in their partner's cards. Matches are discarded. Play is over when there are no more matches. The winner has fewer cards.	Students identify equivalent fractions. [Number and Numeration Goal 5]

Game: *Getting to One*

Skill Practiced: Apply proportional reasoning skills. [Number and Numeration Goal 2]

Modification	Purpose of Modification
Players choose a mystery number between 1 and 20 or a multiple of 5 between 1 and 100.	Students apply proportional reasoning skills. [Number and Numeration Goal 2]
Players use a "What's My Rule?" table to record their guesses (*in* numbers) and calculator displays (*out* numbers). When players figure out the mystery number, they write a rule for the table, for example, ÷ 46.	Students compare and order decimals and identify a rule for a calculator function. [Number and Numeration Goal 6; Patterns, Functions, and Algebra Goal 1]

 **Math Boxes**

Suggestions for using Math Boxes to meet individual needs begin on page 26 of this handbook. There are blank masters for Math Boxes on pages 134–139.

Using Part 3 of the Lessons

Use your professional judgment, along with assessment results, to determine whether the whole class, small groups, or individual students might benefit from these Unit 7 activities. Consider using the Part 3 Planning Master found on page 152 of this handbook to record your plans.

Readiness Activities

Lesson	Activity	Purpose of Activity
7♦1	Make a Fraction Number-Line Poster.	Explore fractions on a number line. [Patterns, Functions, and Algebra Goal 1]
7♦2	Act out "fraction-of-a-collection" situations.	Explore fractions of a collection. [Number and Numeration Goal 2]
7♦3	Examine a deck of regular playing cards and record descriptive information about the cards in the deck.	Explore features of a deck of playing cards. [Data and Chance Goal 4]
7♦4	Use colored squares or tiles to build rectangles with colors in the same fractional parts.	Explore the relationship of fractional parts to the whole or ONE. [Number and Numeration Goal 2]
7♦5	Read *Full House: An Invitation to Fractions,* and write number models for fraction addition.	Explore the addition of fractions with like denominators. [Operations and Computation Goal 5]
7♦6	Find equivalent fractions by matching fractional parts of circles	Explore equivalent names for fractions. [Number and Numeration Goal 5]
7♦7	Identify equivalent fractions using the Fraction Number-Line Poster.	Explore equivalent names for fractions. [Number and Numeration Goal 5]
7♦8	Make base-10 block designs and record their values as fractions and decimals.	Explore representing fractions and decimals on a base-10 grid. [Number and Numeration Goal 2]
7♦9	Sort fractions represented with area and number-line models into groups according to relative size.	Explore comparing fractions. [Number and Numeration Goal 6]
7♦11	Divide circles into equal parts and color specified fractions.	Explore fractional parts of regions. [Number and Numeration Goal 2]
7♦12	Model fractions and their percent equivalents using base-10 blocks.	Explore renaming fractions as percents. [Number and Numeration Goal 5]
7♦12a	Use calculators to skip count by unit fractions.	Explore a fraction $\frac{a}{b}$ as a multiple of $\frac{1}{b}$. [Number and Numeration Goal 3]

English Language Learners Support Activities

Lesson	Activity	Purpose of Activity
7•1	Add *numerator* and *denominator* to the Math Word Bank.	Make connections between new terms and terms students know; use visual models to represent the terms. [Number and Numeration Goal 2]
7•3	Discuss and display the words *impossible, maybe,* and *certain* on a bulletin board or poster.	Use a student-made poster as a visual reference for new terms. [Data and Chance Goal 3]
7•11	Add *fair* and *equal chance* to the Math Word Bank.	Make connections between new terms and terms students know; use visual models to represent the terms. [Data and Chance Goal 3]
7•12	Add *predicted* and *actual* to the Math Word Bank.	Make connections between new terms and terms students know; use visual models to represent the terms. [Data and Chance Goal 4]

Enrichment Activities

Lesson	Activity	Purpose of Activity
7•1	Construct an equilateral triangle and divide it into six equal parts.	Apply understanding of fractions as equal parts of a whole. [Number and Numeration Goal 2]
7•1	Make a design with pattern blocks and name fractional parts of the design.	Apply understanding of the whole. [Number and Numeration Goal 2]
7•2	Solve hiking-trail problems.	Apply understanding of fraction-of situations. [Number and Numeration Goal 2]
7•3	Conduct and analyze a playing-card experiment.	Explore expected and actual results of an experiment. [Data and Chance Goal 4]
7•4	Determine the fractional value of each tangram piece.	Explore the concept of fractional parts of a region. [Number and Numeration Goal 2]
7•4	Write fraction addition number stories.	Apply understanding of addition of fractions. [Operations and Computation Goal 5]
7•6	Model equivalencies of fractions using a clock face.	Explore equivalent fractions. [Number and Numeration Goal 5]
7•7	Investigate how early Egyptians represented fractions as a sum of unit fractions.	Apply understanding of fraction addition and equivalent fractions. [Number and Numeration Goal 5]
7•8	Represent fractions on a base-10 grid and find the percent and decimal equivalents.	Explore fraction, decimal, and percent equivalencies. [Number and Numeration Goal 2]
7•8	Use fractions to design a cap rack.	Apply understanding of fractions with different denominators. [Number and Numeration Goal 5]
7•9	Use the digits 1–9 to make fractions according to relational clues.	Apply understanding of comparing fractions. [Number and Numeration Goal 6]
7•10	Play *Getting to One*.	Apply proportional-reasoning skills and understanding of the concept of ONE. [Number and Numeration Goal 2]
7•10	Use clues to determine how a candy bar was divided.	Apply understanding of the concept of ONE. [Number and Numeration Goal 2]
7•11	Try experiments from *Do You Wanna Bet? Your Chance to Find Out About Probability*.	Apply understanding of probability. [Data and Chance Goal 4]
7•12	Combine results for a cube-drop experiment and compare to predicted results.	Explore the effect of sample size on actual results. [Data and Chance Goal 4]
7•12a	Use different models to multiply a fraction by a whole number.	Explore multiplying a fraction by a whole number. [Operations and Computation Goal 7]

Extra Practice Activities

Lesson	Activity	Purpose of Activity
7◆2	Play *Fraction Of.*	Practice identifying fractions of collections. [Number and Numeration Goal 2]
7◆3	Play *Grab Bag.*	Practice calculating the probability of events. [Data and Chance Goal 4]
7◆5	Solve Frames-and-Arrows problems with fraction rules.	Practice adding and subtracting fractions. [Operations and Computation Goal 5]
7◆6	Play *Fraction Match.*	Practice identifying equivalent fractions. [Number and Numeration Goal 5]
7◆7	Complete name-collection boxes for fractions.	Practice finding equivalent names for fractions. [Number and Numeration Goal 5]
7◆7	Find equivalent fractions with *5-Minute Math* activities.	Practice with equivalent fractions. [Number and Numeration Goal 5]
7◆8	Take a 50-facts test.	Practice multiplication facts. [Operations and Computation Goal 3]
7◆9	Play *Fraction Top-It.*	Practice comparing and ordering fractions. [Number and Numeration Goal 6]
7◆12	Solve *5-Minute Math* probability problems.	Practice with probability. [Data and Chance Goals 3 and 4]
7◆12a	Solve *5-Minute Math* problems about multiplying fractions by whole numbers.	Practice multiplying fractions by whole numbers. [Operations and Computation Goal 7]

Looking at Grade-Level Goals

Everyday Mathematics develops concepts and skills over time. Below is a chart showing where the Grade-Level Goals emphasized in this unit are addressed throughout the year. Use the chart to help you determine which Maintaining Concepts and Skills activities on page 96 to utilize to ensure that students continue working toward these Grade-Level Goals.

Legend:
- ● Grade-Level Goal is taught.
- ◐ Grade-Level Goal is practiced and applied.
- ○ Grade-Level Goal is not a focus.

Grade-Level Goals Emphasized in Unit 7

Goal	1	2	3	4	5	6	7	8	9	10	11	12
[Number and Numeration Goal 2] Read, write, and model fractions; solve problems involving fractional parts of a region or a collection; describe and explain strategies used; given a fractional part of a region or a collection, identify the unit whole.	○	○	●	●	○	◐	◐	◐	◐	◐	○	◐
[Number and Numeration Goal 5] Use numerical expressions to find and represent equivalent names for fractions and decimals; use and explain a multiplication rule to find equivalent fractions; rename fourths, fifths, tenths, and hundredths as decimals and percents.	○	○	○	●	○	◐	◐	◐	◐	◐	◐	◐
[Number and Numeration Goal 6] Compare and order whole numbers up to 1,000,000,000 and decimals through thousandths; compare and order integers between −100 and 0; use area models, benchmark fractions, and analyses of numerators and denominators to compare and order fractions.	●	◐	◐	●	◐	◐	●	◐	●	◐	◐	◐
[Operations and Computation Goal 4] Use manipulatives, mental arithmetic, paper-and-pencil algorithms and models, and calculators to solve problems involving the multiplication of multidigit whole numbers by 2-digit whole numbers and the division of multidigit whole numbers by 1-digit whole numbers; describe the strategies used and explain how they work.	○	○	●	●	●	●	●	●	●	◐	◐	◐
[Operations and Computation Goal 5] Use manipulatives, mental arithmetic, and calculators to solve problems involving the addition and subtraction of fractions and mixed numbers; describe the strategies used.	○	○	○	○	○	◐	●	○	○	◐	○	◐
[Data and Chance Goal 3] Describe events using *certain, very likely, likely, unlikely, very unlikely, impossible,* and other basic probability terms; use *more likely, equally likely, same chance, 50-50, less likely,* and other basic probability terms to compare events; explain the choice of language.	○	○	○	○	○	●	○	○	○	○	○	◐
[Data and Chance Goal 4] Predict the outcomes of experiments and test the predictions using manipulatives; summarize the results and use them to predict future events; express the probability of an event as a fraction.	○	○	○	○	○	○	●	○	○	○	○	◐
[Geometry Goal 2] Describe, compare, and classify plane and solid figures, including polygons, circles, spheres, cylinders, rectangular prisms, cones, cubes, and pyramids, using appropriate geometric terms including *vertex, base, face, edge,* and *congruent.*	○	◐	○	○	○	◐	◐	◐	○	◐	○	○
[Patterns, Functions, and Algebra Goal 1] Extend, describe, and create numeric patterns; describe rules for patterns and use them to solve problems; use words and symbols to describe and write rules for functions that involve the four basic arithmetic operations and use those rules to solve problems.	●	◐	●	○	◐	◐	◐	◐	◐	◐	◐	◐

Maintaining Concepts and Skills

Some of the goals addressed in this unit will be addressed again in later units. Those goals marked with an asterisk (*) are addressed in future units only as practice and application. Here are several suggestions for maintaining concepts and skills until goals are revisited.

Number and Numeration Goal 5

◆ Have students play *Fraction Match*.

◆ Have students identify equivalent fractions using a Fraction Number-Line Poster. See the Readiness activity in Lesson 7-7 for more information.

◆ Use the Name-Collection Boxes master on page 145 of this handbook to create name-collection boxes for equivalent fractions.

Operations and Computation Goal 5*

◆ Have students play *Fraction Top-It* by drawing two cards on each turn and comparing their sums.

◆ Use Frames-and-Arrows masters A and B on pages 142 and 143 of this handbook to create practice problems.

Data and Chance Goal 3*

◆ Have students play *Chances Are*.

◆ Have students discuss the likelihood of everyday events using the language of chance.

Data and Chance Goal 4*

◆ Have students explore probability in the context of a regular deck of playing cards. See the Readiness activity in Lesson 7-3 for more information.

Assessment

See page 102 in the *Assessment Handbook* for modifications to the written portion of the Unit 7 Progress Check.

Additionally, see pages 103–107 for modifications to the open-response task and selected student work samples.

Activities and Ideas for Differentiation

In this unit, students review perimeter and area concepts and develop formulas as mathematical models for area. This section summarizes opportunities for supporting multiple learning styles and ability levels. Use these suggestions to develop a differentiation plan for Unit 8.

Part 1 Activities That Support Differentiation

Below are examples of Unit 8 activities that highlight some of the general instructional strategies that are hallmarks of a differentiated classroom. These strategies will help you support, emphasize, and enhance lesson content to make sure all of your students are engaged in the mathematics at the highest possible level. For more information about general differentiation strategies that accommodate the diverse needs of today's classrooms, see the essay on pages 8–16 of this handbook.

Lesson	Activity	Strategy
8◆2	Students make a scale drawing of the floor plan for their classroom.	Making connections to everyday life
8◆3	Students count squares to estimate the area of polygons.	Modeling visually
8◆4	Students make a square foot from square inches to explore square units.	Modeling physically; using a visual reference
8◆6	Students use straws and connectors to construct a parallelogram.	Modeling concretely
8◆6	Students cut apart and reform a parallelogram to explore the relationship between the areas of a parallelogram and a rectangle with the same base and height.	Modeling physically
8◆7	Students manipulate paper triangles to explore the relationship between the areas of a triangle and a rectangle with the same base and height.	Modeling physically

Vocabulary Development

The list below identifies the Key Vocabulary terms from this unit. The lesson in which each term is defined is indicated next to the term. Some of these terms or their homophones are used outside of mathematics. Consider adding other words as appropriate for developing understanding of the context of the lessons.

Lessons include suggestions for helping English language learners understand and develop vocabulary. For more information, see pages 17–19 of this handbook.

Key Vocabulary

*area **8◆3**	perimeter **8◆1**	square units **8◆3**
*†base **8◆5**	perpendicular **8◆6**	time-and-motion study (†time) **8◆1**
equilateral triangle **8◆7**	right triangle **8◆7**	*variable **8◆5**
*formula **8◆5**	rough floor plan (†rough) **8◆2**	width **8◆5**
height **8◆5**	scalene triangle **8◆2**	work triangle (*work) **8◆1**
isosceles triangle **8◆7**	*scale **8◆2**	
length **8◆5**	scale drawing **8◆2**	

★ Discuss the everyday and mathematical meanings of the words that are marked with an asterisk.

† For words marked with a dagger, write the words and their homophones on the board. For example, *base* and *bass; rough* and *ruff;* and *time* and *thyme.* Discuss and clarify the meaning of each.

◆ As each word is introduced in the lesson, write the word on the board and discuss its meaning.

◆ List the words on a Math Word Wall for students to see. As each word is introduced in the lesson, add a picture next to the word on the Word Wall.

◆ Use the vocabulary words regularly when teaching lessons, and encourage students to use the words in their discussions.

 **Games**

Below are suggested Unit 8 game adaptations. For more information about implementing games in a differentiated classroom, see pages 20–25 of this handbook.

Game: *Fraction Top-It*

Skill Practiced: Compare and order fractions. [Number and Numeration Goal 6]

Modification	Purpose of Modification
Players use the area side of the cards and compare the areas to decide whose fraction is larger.	Students compare fractions using area models. [Number and Numeration Goal 6]
The player with the larger fraction receives a bonus point for telling whether the total of the fractions is greater than or less than 1 (greater than or less than 2 if there are three or four players). They can check with a calculator or by converting to decimals and adding.	Students compare and add fractions. [Number and Numeration Goal 6; Operations and Computation Goal 5]

Game: *Rugs and Fences*

Skill Practiced: Calculate the areas and perimeters of polygons. [Measurement and Reference Frames Goal 2]

Modification	Purpose of Modification
Players calculate either area or perimeter on every turn. The "choice" cards become "double" cards, if the player is correct, the score is doubled.	Students calculate the area or perimeter of polygons. [Measurement and Reference Frames Goal 2]
Players calculate both the area and the perimeter for each polygon. Their score is the combined total. The "choice" cards become "double" cards, if the player is correct, the score is doubled.	Students calculate the area and perimeter of polygons. [Measurement and Reference Frames Goal 2]

Game: *Grab Bag*

Skill Practiced: Calculate the probabilities of events. [Data and Chance Goal 4]

Modification	Purpose of Modification
Players receive 5 bonus points if they can describe the probability of the event using a basic probability term.	Students describe events using probability terms. [Data and Chance Goal 3]
Players receive 5 bonus points if they name their probability in simplest form.	Students calculate the probabilities of events and reduce fractions to their simplest form. [Number and Numeration Goal 5; Data and Chance Goal 4]

 **Math Boxes**

Suggestions for using Math Boxes to meet individual needs begin on page 26 of this handbook. There are blank masters for Math Boxes on pages 134–139.

Using Part 3 of the Lessons

Use your professional judgment, along with assessment results, to determine whether the whole class, small groups, or individual students might benefit from these Unit 8 activities. Consider using the Part 3 Planning Master found on page 152 of this handbook to record your plans.

Readiness Activities

Lesson	Activity	Purpose of Activity
8◆1	Construct rectangles and squares of a given perimeter on a geoboard.	Explore the concept of perimeter. [Measurement and Reference Frames Goal 2]
8◆2	Measure the lengths of objects to the nearest foot using a "foot-long foot."	Explore measuring linear distance. [Measurement and Reference Frames Goal 1]
8◆3	Make squares on a geoboard and find their areas.	Explore the area of a polygon. [Measurement and Reference Frames Goal 2]
8◆4	Determine the area of an irregular region using square stick-on notes as the unit of measure.	Explore the concept of area as the number of unit squares and fractions of unit squares needed to cover a surface. [Measurement and Reference Frames Goal 2]
8◆5	Build rectangles using square pattern blocks, find the areas, and discuss patterns.	Gain experience finding the areas of rectangles. [Measurement and Reference Frames Goal 2]
8◆8	Compare and describe relationships among the areas of shapes.	Explore area comparisons. [Measurement and Reference Frames Goal 2]

English Language Learners Support Activities

Lesson	Activity	Purpose of Activity
8◆2	Brainstorm uses of the word *scale*.	Clarify the mathematical and everyday uses of the term. [Operations and Computation Goal 7]
8◆3	Add *area* and *square unit* to the Math Word Bank.	Make connections between new terms and terms students know; use visual models to represent the terms. [Measurement and Reference Frames Goal 2]
8◆5	Add *length, width, base,* and *height* to the Math Word Bank.	Make connections between new terms and terms students know; use visual models to represent the terms. [Measurement and Reference Frames Goal 2]

Enrichment Activities

Lesson	Activity	Purpose of Activity
8•1	Make polygons with different perimeters using a set of pattern blocks.	Apply understanding of perimeter. [Measurement and Reference Frames Goal 2]
8•2	Make scale drawings of a bedroom and bedroom furniture.	Apply understanding of scale drawings. [Measurement and Reference Frames Goal 1; Operations and Computation Goal 7]
8•3	Make different polygons with the same area on a geoboard.	Apply understanding of area. [Measurement and Reference Frames Goal 2]
8•5	Find the areas of different parts of a tennis court.	Apply understanding of the formula for the area of a rectangle. [Measurement and Reference Frames Goal 2]
8•5	Explore the relationship between the perimeter and area of a rectangle using a string loop and a 1-inch grid.	Apply understanding of perimeter and area. [Measurement and Reference Frames Goal 2]
8•6	Construct parallelograms and perpendicular line segments with a compass and a straightedge.	Apply understanding of the properties of parallelograms. [Geometry Goal 2]
8•6	Explore area and perimeter problems by combining different shapes.	Apply understanding of area and perimeter. [Measurement and Reference Frames Goal 2]
8•7	Compare areas of a rhombus and a hexagon.	Apply understanding of area. [Measurement and Reference Frames Goal 2]
8•7	Find the area and perimeter of a nonregular hexagon.	Apply understanding of area formulas. [Measurement and Reference Frames Goal 2]
8•8	Calculate gravitational pull using patterns in a table of weights.	Apply understanding of comparison strategies. [Operations and Computation Goal 4]
8•8	Explore relationships between the perimeters and areas of similar figures.	Apply understanding of area comparisons. [Measurement and Reference Frames Goal 2]

Extra Practice Activities

Lesson	Activity	Purpose of Activity
8•1	Solve *5-Minute Math* problems involving perimeter.	Practice finding perimeters. [Measurement and Reference Frames Goal 2]
8•2	Take a 50-facts test.	Practice multiplication facts. [Operations and Computation Goal 3]
8•3	Make polygons on geoboards and find their areas.	Practice finding the area of polygons. [Measurement and Reference Frames Goal 2]
8•4	Estimate areas of irregular regions using a grid.	Practice finding the area of irregular regions. [Measurement and Reference Frames Goal 2]
8•7	Play *Rugs and Fences*.	Practice calculating the area and perimeter of polygons. [Measurement and Reference Frames Goal 2]

Looking at Grade-Level Goals

Everyday Mathematics develops concepts and skills over time. Below is a chart showing where the Grade-Level Goals emphasized in this unit are addressed throughout the year. Use the chart to help you determine which Maintaining Concepts and Skills activities on page 103 to utilize to ensure that students continue working toward these Grade-Level Goals.

Legend:
- ● Grade-Level Goal is taught.
- ◐ Grade-Level Goal is practiced and applied.
- ○ Grade-Level Goal is not a focus.

Grade-Level Goals Emphasized in Unit 8	Unit 1	2	3	4	5	6	7	8	9	10	11	12
[Operations and Computation Goal 7] Use repeated addition, skip counting, arrays, area, and scaling to model multiplication and division.	○	○	○	◐	◐	●	◐	●	◐	○	◐	○
[Data and Chance Goal 2] Use the maximum, minimum, range, median, mode, and graphs to ask and answer questions, draw conclusions, and make predictions.	◐	●	●	◐	●	●	◐	●	◐	○	○	○
[Measurement and Reference Frames Goal 1] Estimate length with and without tools; measure length to the nearest $\frac{1}{4}$ inch and $\frac{1}{2}$ centimeter; use tools to measure and draw angles; estimate the size of angles without tools.	●	◐	◐	●	◐	◐	◐	●	◐	◐	◐	○
[Measurement and Reference Frames Goal 2] Describe and use strategies to measure the perimeter and area of polygons, to estimate the area of irregular shapes, and to find the volume of rectangular prisms.	○	○	○	◐	○	○	○	●	○	○	●	○
[Measurement and Reference Frames Goal 3] Describe relationships among U.S. customary units of measure and among metric units of measure.	◐	○	○	●	○	◐	◐	●	○	◐	◐	○
[Geometry Goal 1] Identify, draw, and describe points, intersecting and parallel line segments and lines, rays, and right, acute, and obtuse angles.	◐	◐	○	◐	○	◐	◐	●	○	○	◐	○
[Geometry Goal 2] Describe, compare, and classify plane and solid figures, including polygons, circles, spheres, cylinders, rectangular prisms, cones, cubes, and pyramids, using appropriate geometric terms including *vertex, base, face, edge,* and *congruent.*	●	○	○	◐	○	●	●	●	◐	●	●	○

Maintaining Concepts and Skills

Some of the goals addressed in this unit will be addressed again in later units. Those goals marked with an asterisk (*) are only addressed in future units as practice and application. Here are several suggestions for maintaining concepts and skills until they are formally revisited.

Measurement and Reference Frames Goal 1*

◆ Have students play *Angle Tangle*.

◆ Have students measure the length of objects to the nearest foot. See the Readiness activity in Lesson 8-2 for more information.

Measurement and Reference Frames Goal 2

◆ Have students play *Rugs and Fences*.

◆ Have students construct rectangles and squares of a given perimeter on a geoboard. See the Readiness activity in Lesson 8-1 for more information.

◆ Have students explore area using square pattern blocks. See the Readiness activity in Lesson 8-5 for more information.

Measurement and Reference Frames Goal 3*

◆ Have students measure objects in the room in both inches and centimeters and look for a general relationship between the two units; for example, the measures in centimeters are about two and a half times the measures in inches.

◆ Use the "What's My Rule?" master on page 144 of this handbook to create practice problems with rules involving equivalent measures. For example, the *in* numbers are the number of feet and the *out* numbers are the number of inches.

Geometry Goal 2

◆ Have students play *Polygon Pair-Up*.

◆ Have students build polygons with straws and twist-ties and compare the properties of the polygons.

Assessment

See page 110 in the *Assessment Handbook* for modifications to the written portion of the Unit 8 Progress Check.

Additionally, see pages 111–115 for modifications to the open-response task and selected student work samples.

Activities and Ideas for Differentiation

In this unit, students explore the relationships among and converting among fractions, decimals, and percents with a special emphasis on percents. This section summarizes opportunities for supporting multiple learning styles and ability levels. Use these suggestions to develop a differentiation plan for Unit 9.

Part 1 Activities That Support Differentiation

Below are examples of Unit 9 activities that highlight some of the general instructional strategies that are hallmarks of a differentiated classroom. These strategies will help you support, emphasize, and enhance lesson content to make sure all of your students are engaged in the mathematics at the highest possible level. For more information about general differentiation strategies that accommodate the diverse needs of today's classrooms, see the essay on pages 8–16 of this handbook.

Lesson	Activity	Strategy
9•1	Students use a 100-grid representation to explore relationships among fraction, decimal, and percent names for parts of a whole.	Modeling visually
9•2	Students complete a reference table of equivalent fractions, decimals, and percents.	Using organizational tools
9•3	Students use a calculator to explore the process for finding equivalent decimal names for fractions.	Modeling physically
9•5	Students use a calculator to explore the process for finding equivalent decimal and percent names for fractions.	Modeling physically
9•6	Students explore the meaning and use of percents in the context of survey data they have collected in which the sample size differs.	Making connections to everyday life
9•9	Students explore division of decimals through a variety of contexts such as sharing, partitioning, averaging, and calculating fractions or percents of a quantity.	Making connections to everyday life

Vocabulary Development

The list below identifies the Key Vocabulary terms from this unit. The lesson in which each term is defined is indicated next to the term. Some of these terms or their homophones are used outside of mathematics. Consider adding other words as appropriate for developing understanding of the context of the lessons.

Lessons include suggestions for helping English language learners understand and develop vocabulary. For more information, see pages 17–19 of this handbook.

Key Vocabulary

100% box (*box) **9♦1**	list price (*list) **9♦4**	repeating decimal **9♦3**
discount **9♦4**	percent **9♦1**	rural **9♦7**
discounted price **9♦4**	percent of discount **9♦4**	sale price (†sale) **9♦4**
fraction of discount **9♦4**	*rank **9♦7**	terminating decimal **9♦3**
life expectancy **9♦7**	regular price (*regular) **9♦4**	urban **9♦7**

* Discuss the everyday and mathematical meanings of the words that are marked with an asterisk.

† For the word marked with a dagger, write *sale* and its homophone *sail* on the board. Discuss and clarify the meaning of each.

♦ As each word is introduced in the lesson, write the word on the board and discuss its meaning.

♦ List the words on a Math Word Wall for students to see. As each word is introduced in the lesson, add a picture next to the word on the Word Wall.

♦ Use the vocabulary words regularly when teaching lessons, and encourage students to use the words in their discussions.

 **Games**

Below are suggested Unit 9 game adaptations. For more information about implementing games in a differentiated classroom, see pages 20–25 of this handbook.

Game: *Fraction Match*

Skill Practiced: Identify equivalent fractions. [Number and Numeration Goal 5]

Modification	Purpose of Modification
Players use the visual models on the Everything Math Deck fraction cards that correspond to the *Fraction Match* deck.	Students identify equivalent fractions. [Number and Numeration Goal 5]
Players play a card if it is an equivalent fraction or if the denominator is 1 more or 1 less than the target card. Players must correctly tell if the card with a denominator relationship they wish to play has a larger or smaller fraction than the target number.	Students identify equivalent fractions and compare fractions. [Number and Numeration Goals 5 and 6]

Game: *Rugs and Fences*

Skill Practiced: Calculate the area and perimeter of a polygon. [Measurement and Reference Frames Goal 2]

Modification	Purpose of Modification
Players make their own set of "polygon" cards out of quarter-sheets of grid paper. The cards they make are all rectangles.	Students calculate the area and perimeter of rectangles drawn on grid paper. [Measurement and Reference Frames Goal 2]
Players receive a bonus point if they can draw on grid paper a different polygon that has the same perimeter or area as their polygon. Bonus points count as cards at the end of the game.	Students calculate the area and perimeter of polygons and draw different polygons with the same area and perimeter. [Measurement and Reference Frames Goal 2]

Game: *Fraction/Percent Concentration*

Skill Practiced: Develop automaticity with "easy" fraction/percent equivalencies. [Number and Numeration Goal 5]

Modification	Purpose of Modification
Players use only tenths and the equivalent percents.	Students develop automaticity with fraction/percent equivalencies for tenths. [Number and Numeration Goal 5]
Players receive a bonus point if they can name the equivalent decimal for their fraction/percent pairs. At the end of the game, bonus points count as additional cards.	Students develop automaticity with "easy" fraction/decimal/percent equivalencies. [Number and Numeration Goal 5]

 **Math Boxes**

Suggestions for using Math Boxes to meet individual needs begin on page 26 of this handbook. There are blank masters for Math Boxes on pages 134–139.

Using Part 3 of the Lessons

Use your professional judgment, along with assessment results, to determine whether the whole class, small groups, or individual students might benefit from these Unit 9 activities. Consider using the Part 3 Planning Master found on page 152 of this handbook to record your plans.

Readiness Activities

Lesson	Activity	Purpose of Activity
9♦1	Shade 50% of a square in different ways.	Explore equivalent names for percents. [Number and Numeration Goal 5]
9♦2	Identify and use patterns to solve percent problems.	Explore the relationship between fractions and percents. [Number and Numeration Goal 2]
9♦4	Solve "percent-of" problems by using counters and renaming the percents as "easy" fractions.	Explore "percent-of" situations. [Number and Numeration Goal 2]
9♦5	Round percents using a curved number-line model.	Explore rounding percents to the nearest whole number. [Number and Numeration Goal 5]
9♦6	Compare estimates for what fraction of a collection of pattern blocks is red trapezoids.	Explore the comparison of quantities expressed as fractions with unlike denominators. [Number and Numeration Goal 6]
9♦8	Multiply whole numbers and estimate products.	Gain experience with whole-number multiplication and estimating products. [Operations and Computation Goals 4 and 6]
9♦8	Solve multiplication number stories involving money.	Explore multiplication of whole numbers by decimals. [Operations and Computation Goal 4]
9♦9	Divide whole numbers and estimate quotients.	Gain experience with division. [Operations and Computation Goals 4 and 6]
9♦9	Solve division number stories involving money.	Explore division of decimals by whole numbers. [Operations and Computation Goal 4]

English Language Learners Support Activities

Lesson	Activity	Purpose of Activity
9♦1	Create a *Percents* All Around Museum.	Make connections between mathematics and everyday life; discuss new mathematical ideas. [Number and Numeration Goal 5]
9♦4	Look at store catalogs or advertisements and discuss vocabulary such as *regular price, list price, discount, percent of discount, fraction of discount,* and *sale price.*	Clarify the mathematical and everyday uses of the terms. [Number and Numeration Goal 2]
9♦7	Create a poster to illustrate *rural* and *urban.*	Use a student-made poster as a visual reference for new terms. [Data and Chance Goal 1]

Enrichment Activities

Lesson	Activity	Purpose of Activity
9•2	Write, illustrate, and solve "percent-of" number stories.	Apply understanding of fraction and percent equivalencies. [Number and Numeration Goal 2]
9•3	Name "easy" fractions close to decimals.	Apply understanding of decimal and fraction equivalencies. [Number and Numeration Goal 5]
9•4	Solve discount number stories comparing percents of discounts and actual discounts.	Apply understanding of "percent-of" situations. [Number and Numeration Goal 2]
9•6	Graph class survey results using a side-by-side (double) bar graph.	Apply ability to represent data. [Data and Chance Goal 1]
9•7	Color a map to show literacy data.	Apply ability to display and interpret data. [Data and Chance Goals 1 and 2]
9•8	Compare products of mixed numbers and decimals.	Apply understanding of decimal multiplication and decimal/fraction equivalencies. [Operations and Computation Goals 4 and 6]
9•9	Write and solve decimal-division number stories.	Apply understanding of division of decimals. [Operations and Computation Goal 4]

Extra Practice Activities

Lesson	Activity	Purpose of Activity
9•2	Use base-10 blocks and grids to add tenths and hundredths.	Practice adding fractions with 10 and 100 in the denominator. [Number and Numeration Goal 5]
9•2	Find equivalent names for fractions by shading grids.	Practice finding equivalent decimals and percents for fractions. [Number and Numeration Goal 5]
9•3	Identify equivalent names for "easy" fractions.	Practice with "easy" fraction, decimal, and percent equivalencies. [Number and Numeration Goal 5]
9•4	Solve 5-Minute Math problems involving percents.	Practice with the concept of percent. [Number and Numeration Goal 2]
9•5	Use 5-Minute Math activities to convert among fractions, decimals, and percents.	Practice converting among fractions, decimals, and percents. [Number and Numeration Goal 5]
9•6	Take a 50-facts test.	Practice multiplication facts. [Operations and Computation Goal 3]

Looking at Grade-Level Goals

Everyday Mathematics develops concepts and skills over time. Below is a chart showing where the Grade-Level Goals emphasized in this unit are addressed throughout the year. Use the chart to help you determine which Maintaining Concepts and Skills activities on page 110 to utilize to ensure that students continue working toward these Grade-Level Goals.

Legend:
- ● Grade-Level Goal is taught.
- ◐ Grade-Level Goal is practiced and applied.
- ○ Grade-Level Goal is not a focus.

Grade-Level Goals Emphasized in Unit 9	1	2	3	4	5	6	7	8	9	10	11	12
[Number and Numeration Goal 1] Read and write whole numbers up to 1,000,000,000 and decimals through thousandths; identify places in such numbers and the values of the digits in those places; translate between whole numbers and decimals represented in words and in base-10 notation.	◐	●	●	●	●	◐	○	○	◐	○	○	○
[Number and Numeration Goal 2] Read, write, and model fractions; solve problems involving fractional parts of a region or a collection; describe and explain strategies used; given a fractional part of a region or a collection, identify the unit whole.	○	○	○	○	○	●	●	○	●	○	○	●
[Number and Numeration Goal 5] Use numerical expressions to find and represent equivalent names for fractions and decimals; use and explain a multiplication rule to find equivalent fractions; rename fourths, fifths, tenths, and hundredths as decimals and percents.	○	○	○	○	○	●	●	●	●	●	○	●
[Number and Numeration Goal 6] Compare and order whole numbers up to 1,000,000,000 and decimals through thousandths; compare and order integers between −100 and 0; use area models, benchmark fractions, and analyses of numerators and denominators to compare and order fractions.	◐	○	○	○	○	●	●	●	●	○	○	●
[Operations and Computation Goal 6] Make reasonable estimates for whole number and decimal addition and subtraction problems and whole number multiplication and division problems; explain how the estimates were obtained.	○	◐	◐	◐	◐	◐	◐	◐	◐	◐	◐	◐
[Data and Chance Goal 1] Collect and organize data or use given data to create charts, tables, graphs, and line plots.	○	●	○	◐	○	○	○	○	◐	◐	○	●
[Data and Chance Goal 2] Use the maximum, minimum, range, median, mode, and graphs to ask and answer questions, draw conclusions, and make predictions.	◐	●	●	●	◐	○	○	○	◐	◐	○	●
[Patterns, Functions, and Algebra Goal 2] Use conventional notation to write expressions and number sentences using the four basic arithmetic operations; determine whether number sentences are true or false; solve open sentences and explain the solutions; write expressions and number sentences to model number stories.	○	◐	◐	◐	○	○	○	○	◐	○	○	○

Maintaining Concepts and Skills

Those goals marked with an asterisk (*) are addressed in future units only as practice and application. Here are several suggestions for maintaining concepts and skills until goals are revisited.

Number and Numeration Goal 2*

◆ Have students identify and use patterns to solve percent problems. See the Readiness activity in Lesson 9-2 for more information.

◆ Have students solve "percent-of" problems using counters. See the Readiness activity in Lesson 9-4 for more information.

Number and Numeration Goal 5*

◆ Have students play *Fraction Match*.

◆ Have students round percents using a curved number-line model. See the Readiness Activity in Lesson 9-5 for more information.

◆ Have students find equivalent names for fractions by shading grids. See the Extra Practice activity in Lesson 9-2 for more information.

◆ Use the Name-Collection Boxes master on page 145 of this handbook to create name-collection boxes for fractions or mixed numbers.

Number and Numeration Goal 6*

◆ Have students play *Number Top-It* (Decimals).

◆ Have students compare estimates for the fraction of a collection. See the Readiness activity in Lesson 9-6 for more information.

Patterns, Functions, and Algebra Goal 2*

◆ Have students play *Name That Number* and record number sentences with parentheses for their solutions.

Assessment

See page 118 in the *Assessment Handbook* for modifications to the written portion of the Unit 9 Progress Check.

Additionally, see pages 119–123 for modifications to the open-response task and selected student work samples.

Unit 10 Activities and Ideas for Differentiation

In this unit, students explore transformations, or the "motions" of geometric figures, in the context of reflections, rotations, translations, and congruence. This section summarizes opportunities for supporting multiple learning styles and ability levels. Use these suggestions to develop a differentiation plan for Unit 10.

Part 1 Activities That Support Differentiation

Below are examples of Unit 10 activities that highlight some of the general instructional strategies that are hallmarks of a differentiated classroom. These strategies will help you support, emphasize, and enhance lesson content to make sure all of your students are engaged in the mathematics at the highest possible level. For more information about general differentiation strategies that accommodate the diverse needs of today's classrooms, see the essay on pages 8–16 of this handbook.

Lesson	Activity	Strategy
10•1	Students use transparent mirrors to explore translations.	Modeling physically
10•2	Students use transparent mirrors to explore lines of reflection.	Modeling physically
10•3	Students use transparent mirrors to explore properties of reflections and congruent figures.	Modeling physically
10•4	Students identify lines of symmetry in polygons by folding paper models of polygons.	Modeling physically
10•5	Students explore reflections, rotations, and translations in the context of frieze patterns.	Making connections to everyday life
10•6	Students investigate the relationship between positive and negative numbers in the context of reflections.	Building on prior knowledge

Vocabulary Development

The list below identifies the Key Vocabulary terms from this unit. The lesson in which each term is defined is indicated next to the term. Some of these terms or their homophones are used outside of mathematics. Consider adding other words as appropriate for developing understanding of the context of the lessons.

Lessons include suggestions for helping English language learners understand and develop vocabulary. For more information, see pages 17–19 of this handbook.

Key Vocabulary

congruent **10♦1**

*credit **10♦6**

debit **10♦6**

frieze pattern (†frieze) **10♦5**

*image **10♦1**

line of reflection **10♦2**

line of symmetry **10♦4**

opposite (of a number) **10♦6**

preimage **10♦1**

recessed **10♦1**

*reflection **10♦2**

reflection (flip) **10♦5**

rotation (†turn) **10♦5**

rotation (†turn) symmetry **10♦4**

symmetric **10♦4**

*translation (slide) **10♦5**

transparent mirror **10♦1**

* Discuss the everyday and mathematical meanings of the words that are marked with an asterisk.

† For words marked with a dagger, write the words and their homophones on the board. For example, *frieze* and *freeze* and *turn* and *tern*. Discuss and clarify the meaning of each.

◆ As each word is introduced in the lesson, write the word on the board and discuss its meaning.

◆ List the words on a Math Word Wall for students to see. As each word is introduced in the lesson, add a picture next to the word on the Word Wall.

◆ Use the vocabulary words regularly when teaching lessons, and encourage students to use the words in their discussions.

 # Games

Below are suggested Unit 10 game adaptations. For more information about implementing games in a differentiated classroom, see pages 20–25 of this handbook.

Game: *Over and Up Squares*

Skill Practiced: Locate and plot ordered number pairs on a coordinate grid. [Measurement and Reference Frames Goal 4]

Modification	Purpose of Modification
Players roll one die two times. The first roll is the *x*-coordinate and the second roll is the *y*-coordinate.	Students locate and plot specified ordered number pairs on a coordinate grid. [Measurement and Reference Frames Goal 4]
Players draw a 4-quadrant grid on grid paper from −10 to 10 on each axis. They use two sets of 0–10 cards in one color for positive numbers and two sets of cards in another color for negative numbers to generate coordinates.	Students locate and plot ordered number pairs on a 4-quadrant coordinate grid. [Measurement and Reference Frames Goal 4]

Game: *Angle Tangle*

Skill Practiced: Estimate the size of and measure angles. [Measurement and Reference Frames Goal 1]

Modification	Purpose of Modification
Players make angle templates by drawing and cutting out angles of 20°, 30°, 45°, 60°, 90°, 135°, 180°, 225°, and 270°. They select and trace the angle templates as they play the game.	Students draw, estimate the size of, and measure basic angles. [Measurement and Reference Frames Goal 1]
Each player draws an angle. Both players estimate the difference between the two angle measures. The player with the closer estimate scores a point. The winner is the player with the most points.	Students estimate the size of and measure angles. [Measurement and Reference Frames Goal 1]

Game: *Polygon Pair-Up*

Skill Practiced: Identify properties of polygons. [Geometry Goal 2]

Modification	Purpose of Modification
Players make a list of polygon properties. Only Polygon Cards are used. On each turn, they turn over a Polygon Card and win the polygon if they can name one of its properties. They get a bonus point if they can name a second property. For scoring, bonus points count as cards.	Students describe the properties of specified polygons. [Geometry Goal 2]
Players use WILD Property Cards to name a property of their choice. They collect all face-up Polygon Cards with that property. They use a WILD Polygon Card similarly collecting all Property Cards that match the named polygon.	Students match properties of polygons to polygons. [Geometry Goal 2]

Math Boxes

Suggestions for using Math Boxes to meet individual needs begin on page 26 of this handbook. There are blank masters for Math Boxes on pages 134–139.

Using Part 3 of the Lessons

Use your professional judgment, along with assessment results, to determine whether the whole class, small groups, or individual students might benefit from these Unit 10 activities. Consider using the Part 3 Planning Master found on page 152 of this handbook to record your plans.

Readiness Activities

Lesson	Activity	Purpose of Activity
10◆1	Fold a piece of paper in half and cut out a shape that includes the fold as one side. Predict the final shape and unfold to check the prediction.	Explore problems involving spatial visualization. [Geometry Goal 3]
10◆2	Create a paint reflection.	Explore the concept of reflection. [Geometry Goal 3]
10◆3	Create reflections with pattern blocks or centimeter cubes.	Explore reflections. [Geometry Goal 3]
10◆5	Create and continue geometric patterns using pattern blocks.	Explore geometric patterns. [Patterns, Functions, and Algebra Goal 1]
10◆6	Skip count on a calculator.	Explore patterns in negative numbers. [Operations and Computation Goal 2]
10◆6	Add positive and negative numbers by walking on a life-size number line.	Gain experience adding positive and negative numbers. [Operations and Computation Goal 2]

English Language Learners Support Activities

Lesson	Activity	Purpose of Activity
10◆2	Add *reflection* to the Math Word Bank.	Make connections between a new term and terms students know; use a visual model to represent the term. [Geometry Goal 3]
10◆4	Create a *Line Symmetry* Museum.	Make connections between mathematics and everyday life; discuss new mathematical ideas. [Geometry Goal 3]

Enrichment Activities

Lesson	Activity	Purpose of Activity
10◆1	Explore shadows and reflections in the book *Shadows and Reflections*.	Apply understanding of reflections. [Geometry Goal 3]
10◆2	Solve paper-folding puzzles.	Apply understanding of lines of reflection and reflected images. [Geometry Goal 3]
10◆2	Explore reflections and lines of reflection using a virtual manipulative.	Apply understanding of reflections and lines of reflection. [Geometry Goal 3]
10◆3	Construct 3-dimensional cube buildings and their reflections.	Apply understanding of reflections. [Geometry Goal 3]
10◆4	Interpret a cartoon involving line symmetry.	Apply understanding of line symmetry. [Geometry Goal 3]
10◆4	Use pattern blocks to perform and analyze transformations.	Explore rotation or turn symmetry. [Geometry Goal 3]
10◆5	Arrange four straws in 16 different ways.	Apply understanding of congruence, reflections, and rotations. [Geometry Goals 2 and 3]
10◆5	Explore tessellations using a virtual manipulative.	Apply understanding of reflections, rotations, and translations. [Geometry Goal 3]

Extra Practice Activities

Lesson	Activity	Purpose of Activity
10◆3	Solve *5-Minute Math* problems involving reflections.	Practice with reflections. [Geometry Goal 3]
10◆4	Use pattern blocks to make figures with a specified number of lines of symmetry.	Practice with lines of symmetry. [Geometry Goal 3]
10◆5	Create frieze patterns.	Practice making reflections, rotations, and translations. [Geometry Goal 3]

Looking at Grade-Level Goals

Everyday Mathematics develops concepts and skills over time. Below is a chart showing where the Grade-Level Goals emphasized in this unit are addressed throughout the year. Use the chart to help you determine which Maintaining Concepts and Skills activities on page 117 to utilize to ensure that students continue working toward these Grade-Level Goals.

- ● Grade-Level Goal is taught.
- ◐ Grade-Level Goal is practiced and applied.
- ○ Grade-Level Goal is not a focus.

Grade-Level Goals Emphasized in Unit 10	Unit 1	2	3	4	5	6	7	8	9	10	11	12
[Geometry Goal 2] Describe, compare, and classify plane and solid figures, including polygons, circles, spheres, cylinders, rectangular prisms, cones, cubes, and pyramids, using appropriate geometric terms including *vertex, base, face, edge,* and *congruent*.	●	◐	◐	◐	○	●	●	●	○	●	●	○
[Geometry Goal 3] Identify, describe, and sketch examples of reflections; identify and describe examples of translations and rotations.	○	◐	○	◐	○	●	○	○	◐	●	◐	○
[Patterns, Functions, and Algebra Goal 1] Extend, describe, and create numeric patterns; describe rules for patterns and use them to solve problems; use words and symbols to describe and write rules for functions that involve the four basic arithmetic operations and use those rules to solve problems.	◐	◐	◐	○	●	◐	◐	◐	◐	◐	○	○

Maintaining Concepts and Skills

Some of the goals addressed in this unit will be addressed again in later units. Those goals marked with an asterisk (*) are addressed in future units only as practice and application. Here are several suggestions for maintaining concepts and skills until goals are revisited.

Geometry Goal 2

◆ Have students play *Polygon Pair-Up*.

◆ Have students build shapes with straws and twist-ties and compare the shapes.

Geometry Goal 3*

◆ Have students make paint reflections. See the Readiness activity in Lesson 10-2 for more information.

◆ Have students make reflected designs with pattern blocks or centimeter cubes. See the Readiness activity in Lesson 10-3 for more information.

Patterns, Functions, and Algebra Goal 1*

◆ Have students create and continue geometric patterns using pattern blocks. See the Readiness activity in Lesson 10-5 for more information.

◆ Use Frames-and-Arrows masters A and B on pages 142 and 143 of this handbook to create practice problems.

◆ Use the "What's My Rule?" master on page 144 of this handbook to create practice problems.

Assessment

See page 126 in the *Assessment Handbook* for modifications to the written portion of the Unit 10 Progress Check.

Additionally, see pages 127–131 for modifications to the open-response task and selected student work samples.

Unit 11 Activities and Ideas for Differentiation

In this unit, students review and extend their work with properties of 3-dimensional shapes and explore weight and capacity. This section summarizes opportunities for supporting multiple learning styles and ability levels. Use these suggestions to develop a differentiation plan for Unit 11.

Part 1 Activities That Support Differentiation

Below are examples of Unit 11 activities that highlight some of the general instructional strategies that are hallmarks of a differentiated classroom. These strategies will help you support, emphasize, and enhance lesson content to make sure all of your students are engaged in the mathematics at the highest possible level. For more information about general differentiation strategies that accommodate the diverse needs of today's classrooms, see the essay on pages 8–16 of this handbook.

Lesson	Activity	Strategy
11•1	Students use a number line with two scales as a visual reference for converting between grams and ounces.	Using visual references
11•2	Students use geometric-solid models to compare the properties of geometric solids.	Modeling concretely
11•3	Students build 3-dimensional shapes with straws and twist-ties to compare and discuss properties of 3-dimensional shapes.	Modeling concretely
11•4	Students use base-10 longs and metersticks to explore the relationships among metric units of length.	Modeling visually
11•5	Students compute the volume of rectangular prisms using models that show the numbers of cubes for the required dimensions.	Modeling visually
11•6	Students explore subtracting positive and negative numbers in the context of credits and debits for a business.	Making connections to everyday life

Vocabulary Development

The list below identifies the Key Vocabulary terms from this unit. The lesson in which each term is defined is indicated next to the term. Some of these terms or their homophones are used outside of mathematics. Consider adding other words as appropriate for developing understanding of the context of the lessons.

Lessons include suggestions for helping English language learners understand and develop vocabulary. For more information, see pages 17–19 of this handbook.

Key Vocabulary

capacity 11◆7	flat surface (*flat) 11◆2	quart 11◆7
cone 11◆2	*formula 11◆5	rectangular prism 11◆2
congruent 11◆2	gallon 11◆7	regular polyhedron 11◆3
cube 11◆2	geometric solid 11◆2	sphere 11◆2
cubic units 11◆4	gram 11◆1	square pyramid 11◆2
*cup 11◆7	liter 11◆7	surface area 11◆4
curved surface 11◆2	milliliter 11◆7	tetrahedron 11◆3
*cylinder 11◆2	*ounce 11◆1	triangular prism 11◆2
dimensions 11◆4	pint 11◆7	triangular pyramid 11◆3
dodecahedron 11◆3	polyhedron 11◆3	vertex (vertices) 11◆2
*edge 11◆2	*prism 11◆3	*volume 11◆4, 11◆5
*face 11◆2	pyramid 11◆3	

* Discuss the everyday and mathematical meanings of the words that are marked with an asterisk.

◆ As each word is introduced in the lesson, write the word on the board and discuss its meaning.

◆ List the words on a Math Word Wall for students to see. As each word is introduced in the lesson, add a picture next to the word on the Word Wall.

◆ Use the vocabulary words regularly when teaching lessons, and encourage students to use the words in their discussions.

 **Games**

Below are suggested Unit 11 game adaptations. For more information about implementing games in a differentiated classroom, see pages 20–25 of this handbook.

Game: *Credits/Debits Game*

Skill Practiced: Add positive and negative numbers. [Operations and Computation Goal 2]

Modification	Purpose of Modification
Players use a number line to model the addition and subtraction problems.	Students add positive and negative numbers on a number line. [Operations and Computation Goal 2]
Players record an addition number sentence for each round of the game.	Students add positive and negative numbers, and record number sentences. [Operations and Computation Goal 2; Patterns, Functions, and Algebra Goal 2]

Game: *Credits/Debits Game* (Advanced Version)

Skill Practiced: Add and subtract positive and negative numbers. [Operations and Computation Goal 2]

Modification	Purpose of Modification
Players use number cards 1–5 (two of each in two colors of each number). They have 20 each of $1 bills, $10 bills, and $1 IOUs made out of paper or index cards to model the problems.	Students add and subtract positive and negative numbers. [Operations and Computation Goal 2]
Players record addition or subtraction number sentences for each round of the game.	Students add and subtract positive and negative numbers, and record number sentences. [Operations and Computation Goal 2; Patterns, Functions, and Algebra Goal 2]

Game: *Chances Are*

Skill Practiced: Use basic probability terms to describe the likelihood of events. [Data and Chance Goal 3]

Modification	Purpose of Modification
Players only use dice and spinner Event Cards.	Students use basic probability terms to describe the likelihood of dice and spinner events. [Data and Chance Goal 3]
Players receive a bonus point if they determine the expected probability of the event on any Event Cards that they match. The game ends when the deck is gone. Bonus points count as cards. The player with the most cards wins.	Students use basic probability terms to describe the likelihood of events, and they determine the expected probability. [Data and Chance Goals 3 and 4]

Math Boxes

Suggestions for using Math Boxes to meet individual needs begin on page 26 of this handbook. There are blank masters for Math Boxes on pages 134–139.

Using Part 3 of the Lessons

Use your professional judgment, along with assessment results, to determine whether the whole class, small groups, or individual students might benefit from these Unit 11 activities. Consider using the Part 3 Planning Master found on page 152 of this handbook to record your plans.

Readiness Activities

Lesson	Activity	Purpose of Activity
11◆1	Compare the weights of objects and put them in order according to their weights.	Explore the concept of weight. [Measurement and Reference Frames Goal 1]
11◆3	Sort common objects by properties.	Explore attributes of geometric solids. [Geometry Goal 2]
11◆4	Build all possible rectangular prisms, each with a different base, using 24 cubes.	Explore the concept of volume. [Measurement and Reference Frames Goal 2]
11◆5	Build cube stacks and solve spatial-visualization problems.	Explore the representation of 3-dimensional figures with 2-dimensional drawings. [Geometry Goal 3]
11◆6	Model subtraction of positive and negative numbers by walking on a life-size number line.	Explore subtraction of positive and negative numbers. [Operations and Computation Goal 2]
11◆7	Sort containers according to capacity.	Explore capacity. [Measurement and Reference Frames Goal 3]

English Language Learners Support Activities

Lesson	Activity	Purpose of Activity
11◆2	Create a Word Wall of the geometric vocabulary *face, edge, vertex, flat surface,* and *curved surface.*	Make connections between new terms and terms students know; use visual models to represent the terms. [Geometry Goal 2]
11◆4	Add *volume* and *cubic units* to the Math Word Bank.	Make connections between new terms and terms students know; use visual models to represent the terms. [Measurement and Reference Frames Goal 2]
11◆7	Add *capacity* to the Math Word Bank.	Make connections between a new term and terms students know; use visual models to represent the term. [Measurement and Reference Frames Goal 3]

Enrichment Activities

Lesson	Activity	Purpose of Activity
11•2	Use a virtual manipulative to explore the relationship among the number of vertices, faces, and edges of polyhedrons.	Apply ability to describe solid figures. [Geometry Goal 2]
11•3	Find all possible nets that can be folded to form a cube.	Apply understanding of attributes of geometric solids. [Geometry Goal 2]
11•4	Create penticubes and compare their surface areas.	Explore volume and surface area. [Measurement and Reference Frames Goal 2]
11•5	Estimate the volume of a sheet of notebook paper.	Apply understanding of volume. [Measurement and Reference Frames Goal 2]

Extra Practice Activities

Lesson	Activity	Purpose of Activity
11•1	Hold objects and compare their weights to benchmarks; then weigh the objects on a scale.	Practice estimating weights with and without tools. [Measurement and Reference Frames Goal 1]
11•1	Solve *5-Minute Math* problems involving units of weight.	Practice with units of weight. [Measurement and Reference Frames Goal 3]
11•2	Compare geometric solids using a Venn diagram.	Practice comparing the attributes of solid figures. [Geometry Goal 2]
11•3	Take a 50-facts test.	Practice multiplication facts. [Operations and Computation Goal 3]
11•6	Solve *5-Minute Math* problems involving the addition and subtraction of positive and negative numbers.	Practice adding and subtracting positive and negative numbers. [Operations and Computation Goal 2]
11•7	Use *5-Minute Math* activities to convert among customary units of capacity.	Practice with units of capacity. [Measurement and Reference Frames Goal 3]

Looking at Grade-Level Goals

Everyday Mathematics develops concepts and skills over time. Below is a chart showing where the Grade-Level Goals emphasized in this unit are addressed throughout the year. Use the chart to help you determine which Maintaining Concepts and Skills activities on page 124 to utilize to ensure that students continue working toward these Grade-Level Goals.

- ● Grade-Level Goal is taught.
- ◐ Grade-Level Goal is practiced and applied.
- ○ Grade-Level Goal is not a focus.

Grade-Level Goals Emphasized in Unit 11	Unit
	1	2	3	4	5	6	7	8	9	10	11	12
[Measurement and Reference Frames Goal 1] Estimate length with and without tools; measure length to the nearest $\frac{1}{4}$ inch and $\frac{1}{2}$ centimeter; use tools to measure and draw angles; estimate the size of angles without tools.	●	◐	◐	●	◐	●	●	●	◐	◐	◐	○
[Measurement and Reference Frames Goal 2] Describe and use strategies to measure the perimeter and area of polygons, to estimate the area of irregular shapes, and to find the volume of rectangular prisms.	○	◐	○	○	◐	●	●	●	◐	●	●	◐
[Measurement and Reference Frames Goal 3] Describe relationships among U.S. customary units of measure and among metric units of measure.	●	◐	◐	●	◐	●	●	●	◐	○	◐	◐
[Geometry Goal 1] Identify, draw, and describe points, intersecting and parallel line segments and lines, rays, and right, acute, and obtuse angles.	●	◐	◐	◐	◐	●	●	◐	◐	○	○	○
[Geometry Goal 2] Describe, compare, and classify plane and solid figures, including polygons, circles, spheres, cylinders, rectangular prisms, cones, cubes, and pyramids, using appropriate geometric terms including *vertex, base, face, edge,* and *congruent.*	●	◐	◐	◐	◐	●	●	◐	○	●	●	◐
[Patterns, Functions, and Algebra Goal 1] Extend, describe, and create numeric patterns; describe rules for patterns and use them to solve problems; use words and symbols to describe and write rules for functions that involve the four basic arithmetic operations and use those rules to solve problems.	◐	◐	●	○	◐	●	◐	◐	○	◐	◐	◐
[Patterns, Functions, and Algebra Goal 3] Evaluate numeric expressions containing grouping symbols; insert grouping symbols to make number sentences true.	○	○	●	◐	◐	◐	◐	○	◐	○	◐	◐

Maintaining Concepts and Skills

Those goals marked with an asterisk (*) are addressed in Unit 12 only as practice and application. Here are several suggestions for maintaining concepts and skills.

Measurement and Reference Frames Goal 1*

◆ Have students play *Angle Tangle*.

◆ Have students estimate and order weights. See the Readiness activity in Lesson 11-1 for more information.

Measurement and Reference Frames Goal 2*

◆ Have students build rectangular prisms with the same volume. See the Readiness activity in Lesson 11-4 for more information.

Measurement and Reference Frames Goal 3*

◆ Have students put containers in order according to capacity. See the Readiness activity in Lesson 11-7 for more information.

Geometry Goal 2*

◆ Have students play *Polygon Pair-Up*.

◆ Have students sort common objects, which represent geometric solids, by properties. See the Readiness activity in Lesson 11-3 for more information.

Patterns, Functions, and Algebra Goal 1*

◆ Use Frames-and-Arrows masters A and B on pages 142 and 143 of this handbook to create practice problems.

◆ Use the "What's My Rule?" master on page 144 of this handbook to create practice problems.

Assessment

See page 134 in the *Assessment Handbook* for modifications to the written portion of the Unit 11 Progress Check.

Additionally, see pages 135–139 for modifications to the open-response task and selected student work samples.

Unit 12 Activities and Ideas for Differentiation

In this unit, students explore the concept of proportional reasoning through rate problems. This section summarizes opportunities for supporting multiple learning styles and ability levels. Use these suggestions to develop a differentiation plan for Unit 12.

Part 1 Activities That Support Differentiation

Below are examples of Unit 12 activities that highlight some of the general instructional strategies that are hallmarks of a differentiated classroom. These strategies will help you support, emphasize, and enhance lesson content to make sure all of your students are engaged in the mathematics at the highest possible level. For more information about general differentiation strategies that accommodate the diverse needs of today's classrooms, see the essay on pages 8–16 of this handbook.

Lesson	Activity	Strategy
12◆1	Students explore rates in the context of collecting data on and comparing eye-blinking rates.	Making connections to everyday life
12◆2	Students describe strategies for determining rates.	Talking about math
12◆2	Students organize information about rates in rate tables.	Using organizational tools
12◆3	Students explore the meaning of and reasonableness of rates in the context of statistics about what people do in an average lifetime.	Making connections to everyday life
12◆4	Students apply what they have learned about rates in the context of comparison shopping.	Making connections to everyday life
12◆5	Students use calculators to compute and compare complex rates.	Modeling physically

Vocabulary Development

The list below identifies the Key Vocabulary terms from this unit. The lesson in which each term is defined is indicated next to the term. Some of these terms or their homophones are used outside of mathematics. Consider adding other words as appropriate for developing understanding of the context of the lessons.

Lessons include suggestions for helping English language learners understand and develop vocabulary. For more information, see pages 17–19 of this handbook.

Key Vocabulary

comparison shopping **12♦4**

consumer **12♦4**

†per **12♦1**

*products **12♦4**

*rate **12♦1**

rate table (*table) **12♦2**

*services **12♦4**

unit price (*unit) **12♦4**

unit rate **12♦2**

* Discuss the everyday and mathematical meanings of the words that are marked with an asterisk.

† For the word marked with a dagger, write *per* and its homophone *purr* on the board. Discuss and clarify the meaning of each.

♦ As each word is introduced in the lesson, write the word on the board and discuss its meaning.

♦ List the words on a Math Word Wall for students to see. As each word is introduced in the lesson, add a picture next to the word on the Word Wall.

♦ Use the vocabulary words regularly when teaching lessons, and encourage students to use the words in their discussions.

 **Games**

Below are suggested Unit 12 game adaptations. For more information about implementing games in a differentiated classroom, see pages 20–25 of this handbook.

Game: *Credits/Debits Game* (Advanced Version)

Skill Practiced: Add and subtract positive and negative numbers. [Operations and Computation Goal 2]

Modification	Purpose of Modification
Players use a calculator to perform the addition or subtraction.	Students add and subtract positive and negative numbers on a calculator. [Operations and Computation Goal 2]
Players draw two cards on each turn and add or subtract the sum from the "Start" balance.	Students perform multistep addition and subtraction with positive and negative numbers. [Operations and Computation Goal 2]

Game: *Name That Number*

Skill Practiced: Write equivalent names for numbers. [Number and Numeration Goal 4; Patterns, Functions, and Algebra Goal 3]

Modification	Purpose of Modification
Players use only addition and subtraction to make the target number.	Students write equivalent names for numbers using addition and subtraction. [Number and Numeration Goal 4]
Players use at least four cards and two different operations for each solution. Players receive a bonus point if they use all five cards. Bonus points count as cards at the end of the game.	Students write equivalent names for numbers using at least two operations. [Number and Numeration Goal 4]

Game: *Fraction Top-It*

Skill Practiced: Compare and order fractions. [Number and Numeration Goal 6]

Modification	Purpose of Modification
Players use only the cards showing halves, fourths, eighths, and twelfths.	Students compare and order fractions. [Number and Numeration Goal 6]
Players draw two fractions on each turn and compare the sums of their fraction pairs.	Students compare, order, and add fractions. [Number and Numeration Goal 6; Operations and Computation Goal 5]

 **Math Boxes**

Suggestions for using Math Boxes to meet individual needs begin on page 26 of this handbook. There are blank masters for Math Boxes on pages 134–139.

Using Part 3 of the Lessons

Use your professional judgment, along with assessment results, to determine whether the whole class, small groups, or individual students might benefit from these Unit 12 activities. Consider using the Part 3 Planning Master found on page 152 of this handbook to record your plans.

Readiness Activities

Lesson	Activity	Purpose of Activity
12◆1	Analyze the median and mean of spelling-test scores.	Gain experience finding the median and calculating the mean of a data set. [Data and Chance Goal 2]
12◆2	Use situations in *Each Orange Had 8 Slices: A Counting Book* to describe and illustrate rates.	Explore rate situations. [Operations and Computation Goal 7]
12◆4	Use bills and coins to act out dividing the total cost of goods by the number of items to find the unit prices.	Explore unit prices. [Operations and Computation Goal 7]
12◆5	Solve comparison-shopping problems using drawings and coins and bills.	Explore solving rate problems. [Operations and Computation Goal 7]
12◆6	Solve division problems and interpret remainders.	Gain experience with remainders in division problems. [Operations and Computation Goal 4]

English Language Learners Support Activities

Lesson	Activity	Purpose of Activity
12◆1	Create a *Rates* All Around Museum.	Make connections between mathematics and everyday life; discuss new mathematical ideas. [Operations and Computation Goal 7]
12◆3	Analyze *life expectancy* and *average lifetime* data.	Clarify the mathematical and everyday uses of the terms. [Data and Chance Goal 2]

Enrichment Activities

Lesson	Activity	Purpose of Activity
12◆1	Create a side-by-side (double) bar graph to display eye-blinking rates.	Apply ability to organize and compare data. [Data and Chance Goals 1 and 2]
12◆2	Represent rates with line graphs.	Explore representing rates with line graphs. [Data and Chance Goal 1]
12◆3	Predict and calculate mammal speeds.	Apply understanding of rates. [Operations and Computation Goal 7]
12◆4	Test products; analyze and interpret data.	Apply ability to analyze and interpret data. [Data and Chance Goal 2]
12◆5	Explore how barometric pressure can be used to determine elevation.	Apply understanding of rates. [Operations and Computation Goal 7]

Extra Practice Activities

Lesson	Activity	Purpose of Activity
12◆2	Solve rate problems using rate tables.	Practice solving rate problems. [Operations and Computation Goal 7]
12◆5	Solve *5-Minute Math* problems involving rates.	Practice calculating rates. [Operations and Computation Goal 7]
12◆6	Take a 50-facts test and reflect on individual progress.	Practice multiplication facts. [Operations and Computation Goal 3]
12◆6	Solve *5-Minute Math* problems involving the mean and median of a data set.	Practice finding the mean and median of a data set. [Data and Chance Goal 2]

Looking at Grade-Level Goals

Everyday Mathematics develops concepts and skills over time. Below is a chart showing where the Grade-Level Goals emphasized in this unit are addressed throughout the year. Use the chart to help you determine which Maintaining Concepts and Skills activities on page 131 to utilize to ensure that students continue working toward these Grade-Level Goals.

Legend:
- ● Grade-Level Goal is taught.
- ◐ Grade-Level Goal is practiced and applied.
- ○ Grade-Level Goal is not a focus.

Grade-Level Goals Emphasized in Unit 12	1	2	3	4	5	6	7	8	9	10	11	12
[Number and Numeration Goal 3] Find multiples of whole numbers less than 10; identify prime and composite numbers; find whole-number factors of numbers.	○	◐	◐	◐	●	●	◐	◐	◐	◐	◐	○
[Operations and Computation Goal 3] Demonstrate automaticity with multiplication facts through 10 * 10 and proficiency with related division facts; use basic facts to compute fact extensions such as 30 * 60.	○	●	●	●	●	●	◐	○	○	○	○	○
[Operations and Computation Goal 4] Use manipulatives, mental arithmetic, paper-and-pencil algorithms and models, and calculators to solve problems involving the multiplication of multidigit whole numbers and the division of multidigit whole numbers by 1-digit whole numbers; describe the strategies used and explain how they work.	○	○	○	○	◐	●	●	◐	◐	◐	○	◐
[Operations and Computation Goal 6] Make reasonable estimates for whole number and decimal addition and subtraction problems and whole number multiplication and division problems; explain how the estimates were obtained.	○	●	●	●	●	◐	○	◐	●	○	○	◐
[Operations and Computation Goal 7] Use repeated addition, skip counting, arrays, area, and scaling to model multiplication and division.	○	○	◐	◐	◐	◐	◐	●	◐	○	○	◐
[Data and Chance Goal 2] Use the maximum, minimum, range, median, mode, and graphs to ask and answer questions, draw conclusions, and make predictions.	◐	●	●	●	◐	●	●	◐	◐	◐	◐	◐
[Patterns, Functions, and Algebra Goal 1] Extend, describe, and create numeric patterns; describe rules for patterns and use them to solve problems; use words and symbols to describe and write rules for functions that involve the four basic arithmetic operations and use those rules to solve problems.	●	●	●	○	●	◐	◐	◐	◐	◐	◐	○

Maintaining Concepts and Skills

After completing the curriculum, here are several suggestions for maintaining and practicing concepts and skills.

Operations and Computation Goal 4

◆ Have students play *Division Dash*.

◆ Have students solve division problems and interpret remainders. See the Readiness activity in Lesson 12-6 for more information.

Operations and Computation Goal 7

◆ Have students describe and illustrate situations involving rates. See the Readiness activity in Lesson 12-2 for more information.

◆ Have students explore comparison-shopping problems. See the Readiness activity in Lesson 12-5 for more information.

Data and Chance Goal 2

◆ Have students analyze the median and mean of a data set. See the Readiness activity in Lesson 12-1 for more information.

Patterns, Functions, and Algebra Goal 1

◆ Use Frames-and-Arrows masters A and B on pages 142 and 143 of this handbook to create practice problems.

◆ Use the "What's My Rule?" master on page 144 of this handbook to create practice problems.

Assessment

See page 142 in the *Assessment Handbook* for modifications to the written portion of the Unit 12 Progress Check.

Additionally, see pages 143–147 for modifications to the open-response task and selected student work samples.

ASSESSMENT HANDBOOK

Common Core State Standards Edition

Everyday Mathematics

The University of Chicago School Mathematics Project

Grade 4

Masters

The masters listed below provide additional resources that you can customize to meet the needs of a diverse group of learners. The templates, pages 134–151, include additional Math Boxes problem sets, more practice with program routines, and support for language development. Use the Part 3 Planning Master, page 152, to record information about your differentiation plan.

Contents

Math Boxes A

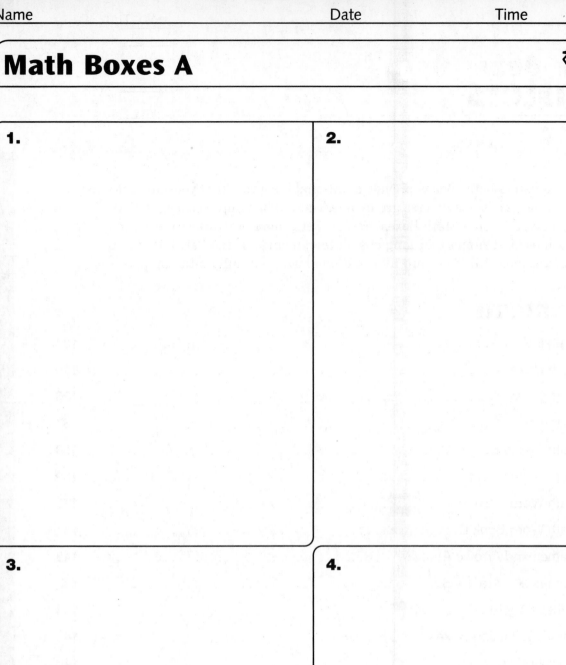

1.

2.

3.

4.

Math Boxes B

1.

2.

3.

4.

5.

6.

Math Boxes C

1.

2.

Rule		in	out

3.

Rule

4. Complete the number-grid puzzles.

Math Boxes D

1. Draw a polygon with an area of
_____ cm².

2. If you spin the spinner _____ times, about how many times would you expect it to land on:

_____? _____ times

_____? _____ times

3. Order the fractions from least to greatest.

▢/▢ , ▢/▢ , ▢/▢ , ▢/▢ , ▢/▢

_____ , _____ , _____ , _____ , _____

4. Write the ordered pair for each point plotted on the coordinate grid.

A (_____,_____) B (_____,_____)

C (_____,_____) D (_____,_____)

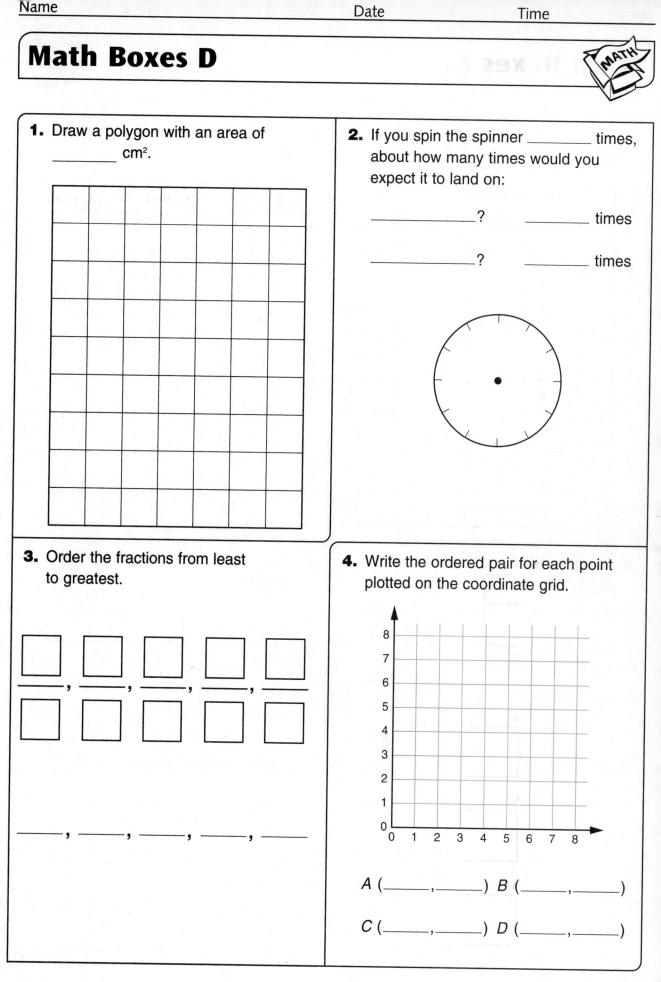

Math Boxes E

1. Draw a polygon with a perimeter of _____ cm.

2. Complete the table with equivalent names.

Fraction	Decimal	Percent

3. Multiply. Use a paper-and-pencil algorithm.

_____ * _____ = _____

4. Fill in the missing numbers on the number lines.

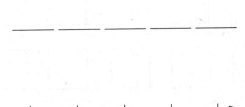

Math Boxes F

1. Solve the open sentences.

    a. _____ + x = _____

       x = _____

    b. _____ − y = _____

       y = _____

2. Use the following data to answer the questions.

_____, _____, _____, _____,

_____, _____, _____

    a. What is the median?

       _____

    b. What is the mean?

       _____

3. Plot the following points on the coordinate grid.

    A (_____,_____) B (_____,_____)

    C (_____,_____) D (_____,_____)

    E (_____,_____) F (_____,_____)

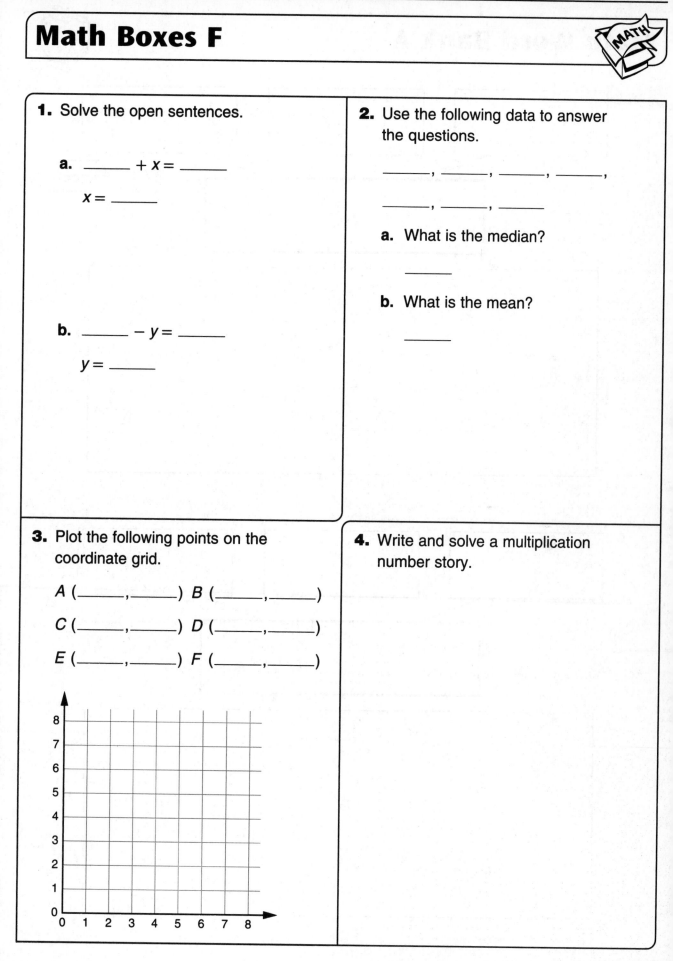

4. Write and solve a multiplication number story.

Math Word Bank A

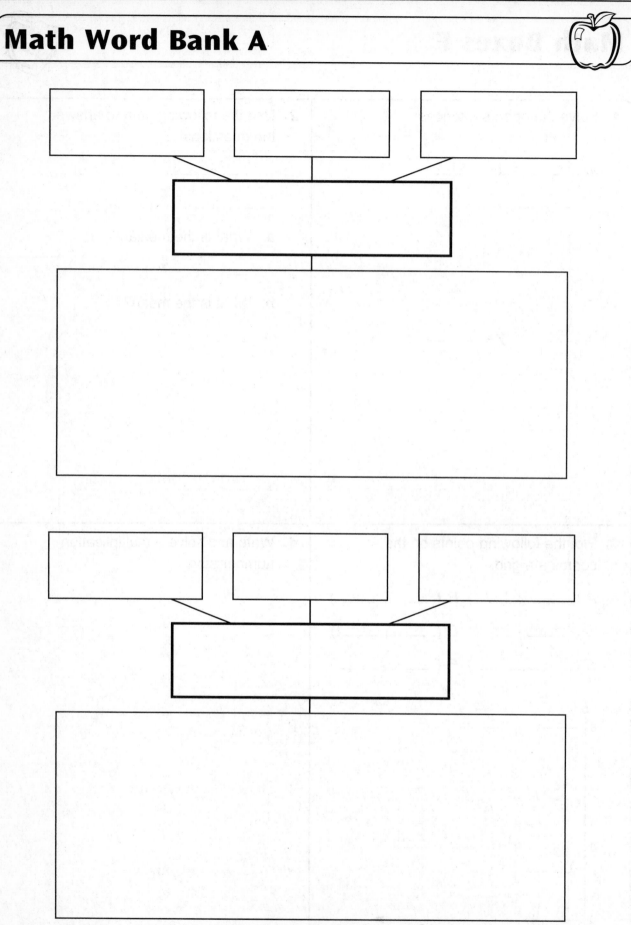

Math Word Bank B

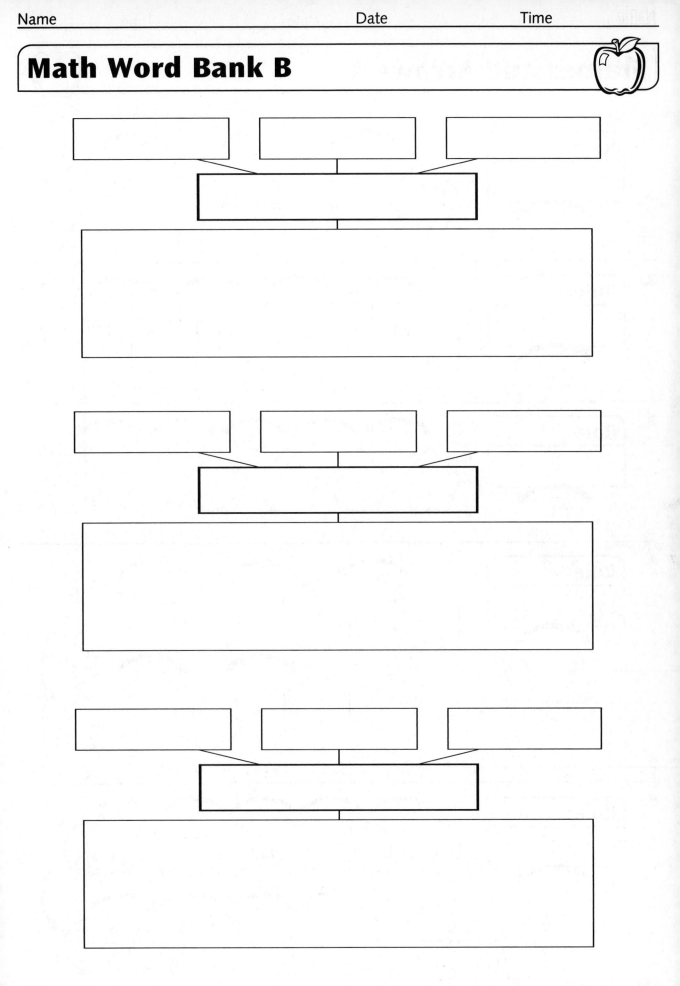

Frames and Arrows A

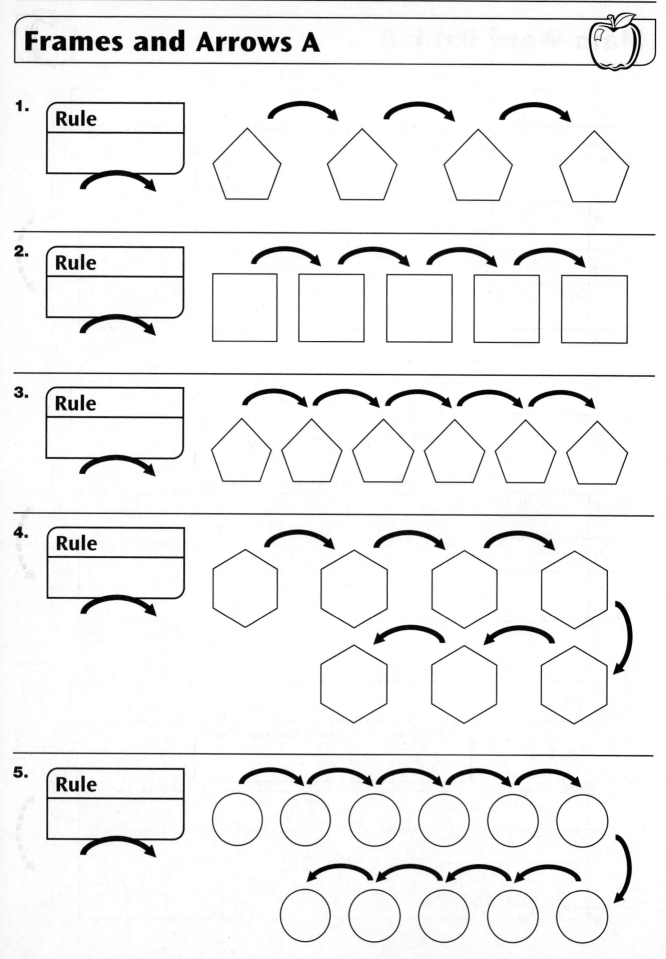

Frames and Arrows B

1.

2.

3.

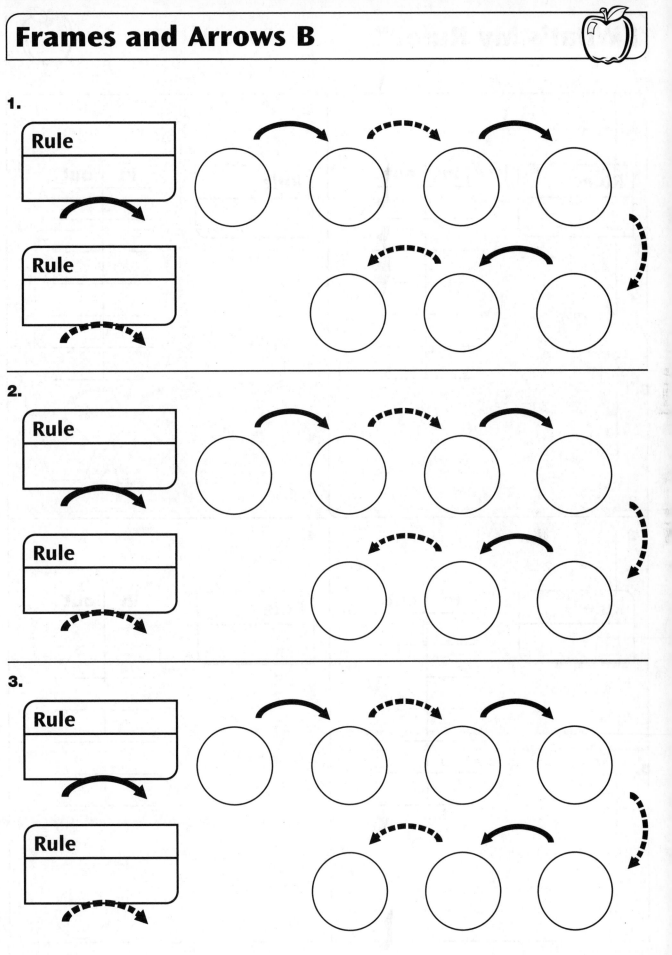

"What's My Rule?"

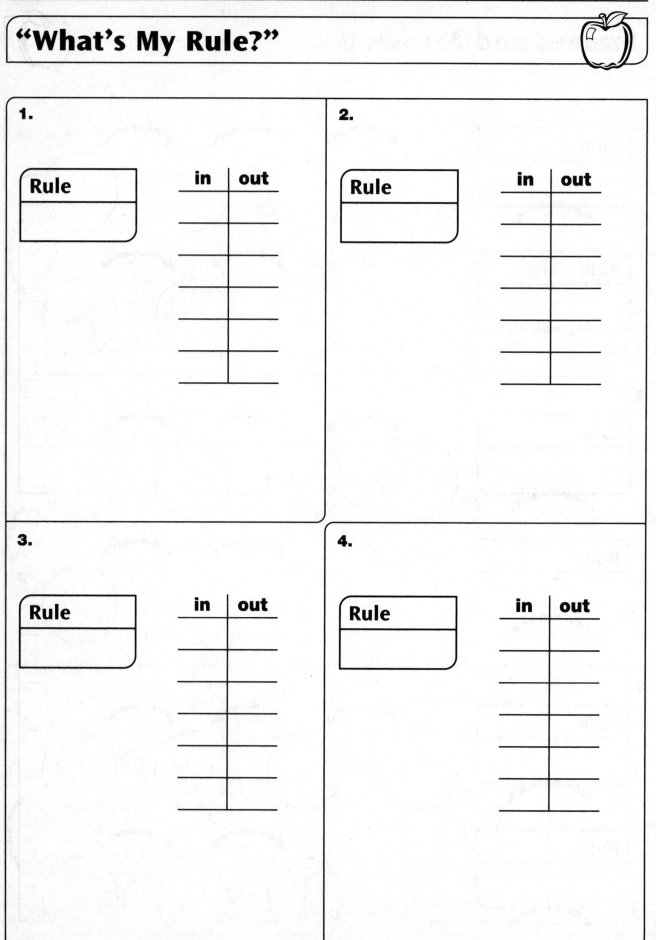

1.

Rule

in	out

2.

Rule

in	out

3.

Rule

in	out

4.

Rule

in	out

Name-Collection Boxes

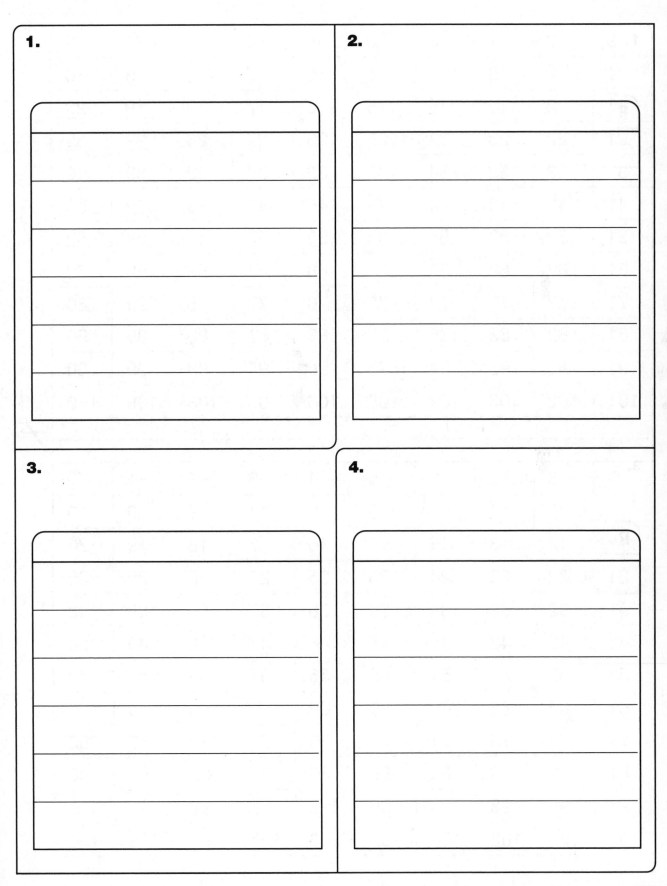

1.

2.

3.

4.

Number Grid

−9	−8	−7	−6	−5	−4	−3	−2	−1	0
1	2	3	4	5	6	7	8	9	10
11	12	13	14	15	16	17	18	19	20
21	22	23	24	25	26	27	28	29	30
31	32	33	34	35	36	37	38	39	40
41	42	43	44	45	46	47	48	49	50
51	52	53	54	55	56	57	58	59	60
61	62	63	64	65	66	67	68	69	70
71	72	73	74	75	76	77	78	79	80
81	82	83	84	85	86	87	88	89	90
91	92	93	94	95	96	97	98	99	100
101	102	103	104	105	106	107	108	109	110

−9	−8	−7	−6	−5	−4	−3	−2	−1	0
1	2	3	4	5	6	7	8	9	10
11	12	13	14	15	16	17	18	19	20
21	22	23	24	25	26	27	28	29	30
31	32	33	34	35	36	37	38	39	40
41	42	43	44	45	46	47	48	49	50
51	52	53	54	55	56	57	58	59	60
61	62	63	64	65	66	67	68	69	70
71	72	73	74	75	76	77	78	79	80
81	82	83	84	85	86	87	88	89	90
91	92	93	94	95	96	97	98	99	100
101	102	103	104	105	106	107	108	109	110

Venn Diagram A

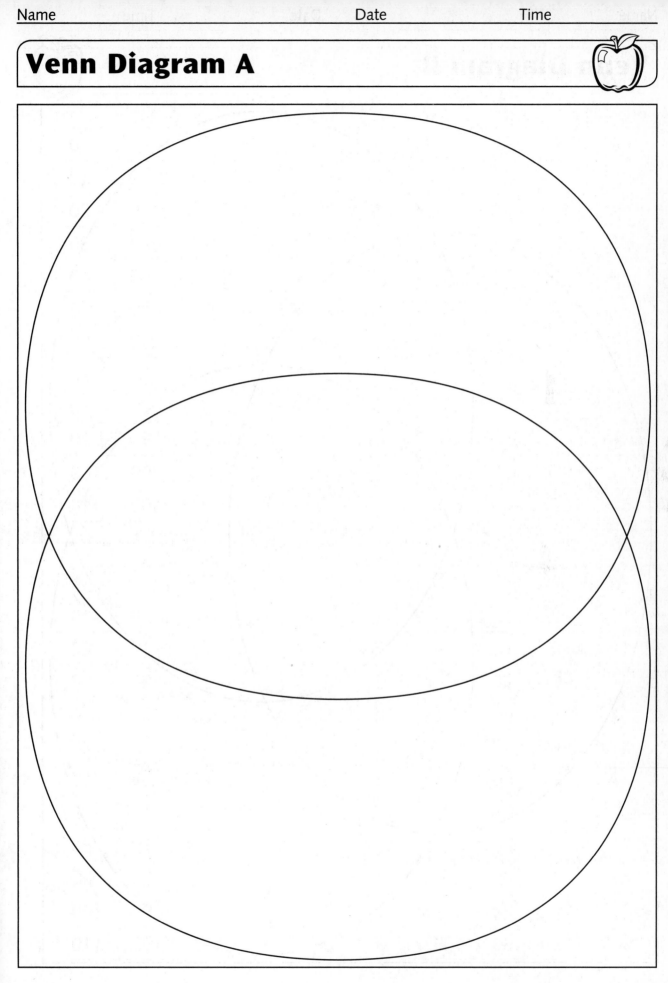

Venn Diagram B

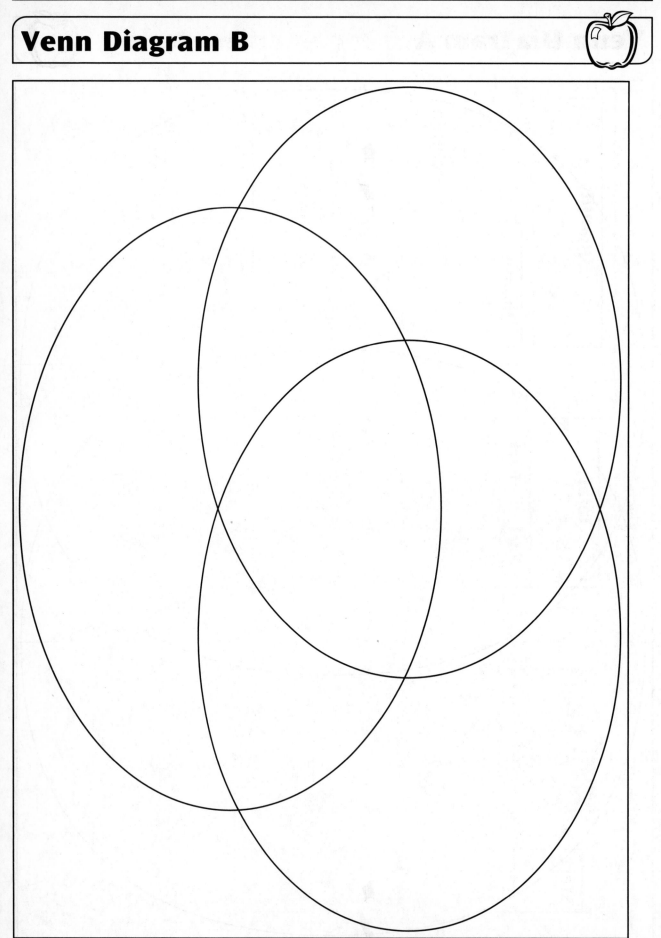

Situation Diagrams for Number Stories

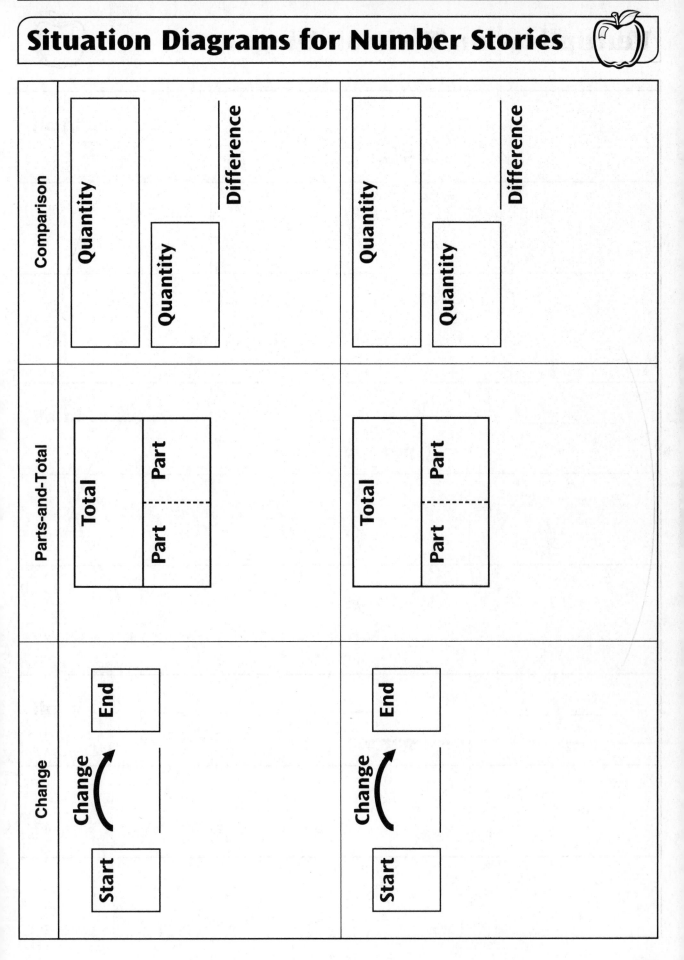

Multiplication/Division Diagrams

_____	_____	_____ **in all**
	per _____	

_____	_____	_____ **in all**
	per _____	

_____	_____	_____ **in all**
	per _____	

Study Link

Part 3 Planning Master

Lesson	Readiness	Enrichment	Extra Practice	ELL Support

Resources

Recommended Reading

Baxter, Juliet A., John Woodward, and Deborah Olson. 2001. "Effects of Reform-Based Mathematics Instruction on Low Achievers in Five Third-Grade Classrooms." *The Elementary School Journal* 101 (5): 529–547.

Garnett, Kate. 1998. "Math Learning Disabilities." LD OnLine. www.ldonline.org (accessed Jan. 19, 2004).

Johnson, Dana T. 2000. "Teaching Mathematics to Gifted Students in a Mixed-Ability Classroom." Reston, Va.: ERIC Clearinghouse on Disabilities and Gifted Education.

Lock, Robin H. 1997. "Adapting Mathematics Instruction in the General Education Classroom for Students with Mathematics Disabilities." LD OnLine. www.ldonline.org (accessed Dec. 15, 2009).

Tomlinson, Carol Ann. 1999. *The Differentiated Classroom: Responding to the Needs of All Learners.* Alexandria, Va.: Association for Supervision & Curriculum Development.

Usiskin, Zalman. 1994. "Individual Differences in the Teaching and Learning of Mathematics." Chicago, Ill.: UCSMP Newsletter 14 (Winter).

Villa, Richard A., and Jacqueline S. Thousand, eds. 2005. *Creating an Inclusive School.* Alexandria, Va.: Association for Supervision & Curriculum Development.

http://everydaymath.uchicago.edu/

References

Gregory, Gayle H. 2003. *Differentiated Instructional Strategies in Practice: Training, Implementation, and Supervision.* Thousand Oaks, Calif.: Corwin Press.

Robertson, Connie, ed. 1998. *Dictionary of Quotations (Wordsworth Reference Series).* 3rd Rev. Edition. Hertfordshire, UK: Wordsworth Editions Ltd.

Tomlinson, Carol Ann. 2003. "Deciding to Teach Them All." *Educational Leadership* 61 (2): 6–11.